Springer Series on

SIGNALS AND COMMUNICATION TECHNOLOGY

Signals and Communication Technology

Multimodal User Interfaces
From Signals to Interaction
D. Tzovaras ISBN 978-3-540-78344-2

Human Factors and Voice Interactive Systems
D. Gardner-Bonneau, H.E. Blanchard (Eds.)
ISBN 978-0-387-25482-1

Wireless Communications
2007 CNIT Thyrrenian Symposium
S. Pupolin (Ed.) ISBN 978-0-387-73824-6

Satellite Communications and Navigation Systems
E. Del Re, M. Ruggieri (Eds.)
ISBN 978-0-387-47522-6

Digital Signal Processing
An Experimental Approach
S. Engelberg ISBN 978-1-84800-118-3

Digital Video and Audio Broadcasting Technology
A Practical Engineering Guide
W. Fischer ISBN 978-3-540-76357-4

Three-Dimensional Television
Capture, Transmission, Display
H.M. Ozaktas, L. Onural (Eds.)
ISBN 978-3-540-72531-2

Foundations and Applications of Sensor Management
A.O. Hero, D. Castañón, D. Cochran, K. Kastella (Eds.) ISBN 978-0-387-27892-6

Digital Signal Processing with Field Programmable Gate Arrays
U. Meyer-Baese ISBN 978-3-540-72612-8

Adaptive Nonlinear System Identification
The Volterra andWiener Model Approaches
T. Ogunfunmi ISBN 978-0-387-26328-1

Continuous-Time Systems
Y.S. Shmaliy ISBN 978-1-4020-6271-1

Blind Speech Separation
S. Makino, T.-W. Lee, H. Sawada (Eds.)
ISBN 978-1-4020-6478-4

Cognitive Radio, Software Defined Radio, and AdaptiveWireless Systems
H. Arslan (Ed.) ISBN 978-1-4020-5541-6

Wireless Network Security
Y. Xiao, D.-Z.Du, X. Shen
ISBN 978-0-387-28040-0

Terrestrial Trunked Radio – TETRA
A Global Security Tool
P. Stavroulakis ISBN 978-3-540-71190-2

Multirate Statistical Signal Processing
O.S. Jahromi ISBN 978-1-4020-5316-0

Wireless Ad Hoc and Sensor Networks
A Cross-Layer Design Perspective
R. Jurdak ISBN 978-0-387-39022-2

Positive Trigonometric Polynomials and Signal Processing Applications
B. Dumitrescu ISBN 978-1-4020-5124-1

Face Biometrics for Personal Identification
Multi-Sensory Multi-Modal Systems
R.I. Hammoud, B.R. Abidi, M.A. Abidi (Eds.)
ISBN 978-3-540-49344-0

Cryptographic Algorithms on Reconfigurable Hardware
F. Rodríguez-Henríquez ISBN 978-0-387-33883-5

Ad-Hoc Networking Towards Seamless Communications
L. Gavrilovska ISBN 978-0-4020-5065-7

Multimedia Database Retrieval
A Human-Centered Approach
P. Muneesawang, L. Guan
ISBN 978-0-387-25627-6

Broadband Fixed Wireless Access
A System Perspective
M. Engels; F. Petre ISBN 978-0-387-33956-6

Acoustic MIMO Signal Processing
Y. Huang, J. Benesty, J. Chen
ISBN 978-3-540-37630-9

Algorithmic Information Theory
Mathematics of Digital Information Processing
P. Seibt ISBN 978-3-540-33218-3

Continuous-Time Signals
Y.S. Shmaliy ISBN 978-1-4020-4817-3

Interactive Video
Algorithms and Technologies
R.I. Hammoud (Ed.) ISBN 978-3-540-33214-5

Handover in DVB-H
Investigation and Analysis
X. Yang ISBN 978-3-540-78629-0

Building Automation:
Communication Systems with EIB/KNX, LON and BACnet
H. Merz, T. Hansemann, C. Hübner
ISBN 978-3-540-88828-4

LDPC Coded Modulations
M. Franceschini, G. Ferrari, R. Raheli
ISBN 978-3-540-69455-7

Michele Franceschini • Gianluigi Ferrari
Riccardo Raheli

LDPC Coded Modulations

Springer

Michele Franceschini
IBM T.J. Watson Research Center
1101 Kitchawan Road, Route 134
Yorktown Heights, NY 10598
USA
michele.fra@gmail.com

Gianluigi Ferrari
Riccardo Raheli
Università di Parma
Dipartimento di Ingegneria
dell'Informazione Viale
G.P. Usberti 181A
43100 Parma
Italy
gianluigi.ferrari@unipr.it
raheli@unipr.it

ISSN 1860-4862
ISBN 978-3-540-69455-7 ISBN 978-3-540-69457-1 (eBook)
DOI 10.1007/978-3-540-69457-1
Springer Dordrecht Heidelberg London New York

Library of Congress Control Number: 2009921992

Cover design: WMXDesign GmbH, Heidelberg, Germany

Printed on acid-free paper

Springer is part of Springer Science+Business Media (www.springer.com)

To Federica and Lucia
who dwell my heart and light my day
- Michele M. Franceschini

To Annuccia and our little baby,
whom we still do not see but already love
- Gianluigi Ferrari

To my family
- Riccardo Raheli

3.4 Decoding Algorithms for LDPC Codes 42
3.4.1 Sum-Product Algorithm 42
3.4.2 Min-Sum Algorithm 49
3.4.3 Alternative Decoding Algorithms 53
3.5 Practical LDPC Code Design from Statistical Description . . . 53
3.6 Encoding Techniques for LDPC Codes 56
3.6.1 Encoding by Matrix Multiplication 56
3.6.2 Recursive Encoding and Structured Codes 58
3.7 Concluding Remarks 59

4 Performance Analysis Techniques 61
4.1 Introduction 61
4.2 Monte Carlo System Simulation 62
4.3 Density Evolution 68
4.4 EXIT Charts 71
4.4.1 EXIT Curves and EXIT Charts 72
4.4.2 SISO Detectors and EXIT Charts 73
4.5 EXIT Charts for LDPC Codes 78
4.6 Concluding Remarks 81

5 LDPC Coded Modulations: Analysis and Design 83
5.1 Introduction 83
5.2 LDPC Coded Modulation Schemes: Basics 85
5.2.1 A Graph-Based Detection/Decoding Approach 85
5.2.2 A "Turbo" Detection/Decoding Approach 86
5.3 Communication System Model 87
5.4 EXIT Chart-Based Performance Analysis 90
5.4.1 The First Iterations 91
5.4.2 The Asymptotic Convergence Region 92
5.5 Upper and Lower Bounds on the BER 93
5.5.1 MI-Based Lower Bound on the BER 93
5.5.2 MI-Based Upper Bound on the BER 94
5.6 Code Design for LDPC Coded Modulations 95
5.6.1 The Need for Optimization 96
5.6.2 Optimizing the EXIT Charts 97
5.7 Code Design for Fast Decoding Convergence 101
5.8 Code Design for a Target BER 110
5.8.1 Target BER-based LDPC Code Design for BEC 112
5.9 Code Design for Dispersive and Partial Response Channels . . 120
5.9.1 System Structure and Analysis 120

Preface

About sixty years ago, Shannon's seminal paper laid the foundations of information theory. In particular, it characterized channel coding as a means for achieving the so-called channel capacity, i.e., to exploit the full information transfer potential of the channel. Since then, both theory and techniques for point-to-point communications have been constantly developed, up to the point that, nowadays, techniques for practically closing the gap to channel capacity exist for several simple channels. This has been made possible by the invention of powerful coding methods, such as turbo codes and low-density parity-check (LDPC) codes. The idea of turbo codes was first published in a conference paper in 1993, where the authors used powerful concatenated codes and an iterative scheme which made possible to effectively—although suboptimally—perform decoding. Since the introduction of turbo codes, a huge amount of resources in the scientific community moved to the investigation of iterative detection and decoding techniques. This eventually led to the rediscovery of LDPC codes in 1995. In fact, LDPC codes were first introduces and analyzed by Robert Gallager in the early Sixties. At that time, the limited available computational power made the use of LDPC codes impractical and prevented scientists from fully understanding their potential. After the introduction of irregular LDPC codes and of practical performance analysis tools in the late Nineties, LDPC codes became the most powerful error correcting codes, enabling reliable transmissions at rates close to the channel capacity for a number of memoryless channels.

LDPC codes were originally designed for binary input memoryless channels. Although the binary input assumption is not really restrictive—LDPC codes can in fact be easily generalized to non-binary input symbols—getting rid of the memoryless assumption is a subtle task. In fact, despite LDPC codes for binary-input memoryless channels admit a decoding algorithm which is asymptotically optimum for increasing codeword lengths—besides being optimum, in a few cases, also for finite codeword lengths—there exists no capacity-achieving coding scheme nor practical optimum decoding algorithm for generic communications channels. Nevertheless, there are practical ways to exploit the properties of LDPC codes to obtain efficient communications also over generic channels. In particular, *LDPC coded modulations* are among the most promising techniques for achieving this goal.

In this book, we will explore the world of LDPC coded modulations intended as a means for using binary LDPC codes for obtaining close-to-capacity performance over generic communication channels. In Chapter 1, we introduce some basic concepts which will be useful in the remainder of the book. In particular, we give a short survey of important concepts, such as mathematical modeling of a communication system, modulation and channel coding, together with a short and self-contained introduction to information theory. In Chapter 2, trellis-based detection strategies for modulations and coded modulation schemes are introduced. These will be basic component blocks of LDPC coded modulations. In Chapter 3, we introduce LDPC codes, describing their structure, representation, best known decoding schemes and encoding techniques. In Chapter 4, we introduce and discuss the most relevant performance analysis techniques for assessing the performance of iterative receivers and their component blocks. In particular, we focus on Monte Carlo methods and extrinsic information transfer (EXIT) charts. In Chapter 5, we introduce the concept of LDPC coded modulations and describe how to apply EXIT charts to analyze their performance. We discuss optimization of LDPC codes for LDPC coded modulations considering some relevant optimization targets such as best power efficiency, minimum number of decoding iterations, and minimum bit error rate (BER). As a particularly relevant case study, we consider code optimization for partial response channels. In Chapter 6, we consider LDPC codes for memoryless channels and the implications of the adopted analysis technique on the structure of LDPC codes in a few interesting cases. The results, which demonstrate a low dependence of optimized LDPC codes on the particular memoryless channel, are used to devise a method for designing multilevel coding schemes using a database of LDPC codes. In Chapter 7, we apply the code design techniques described in Chapter 5 to phase-uncertain communication channels considering LDPC coded differential modulations and obtaining insights on the optimized LDPC code structure. We also describe a low-complexity detection strategy particularly suited for use in an LDPC coded modulation system. In Chapter 8, we draw some final remarks.

Last, but not least, we would like to thank Dr. Christoph Baumann, our Springer Engineering Editor, for expressing his interest and supporting this book from the very beginning, and for his patience in waiting for the final delivery.

Chapter 1

Preliminary Concepts

1.1 Introduction

In this chapter, we will present some preliminary concepts which will be extensively used throughout this book. In Section 1.2, we shortly discuss the modeling process of a (high performance) communication system. In Section 1.3, we focus on the concept of modulation and its role in communication schemes. In Section 1.4, some introductory material on error correcting codes is presented, with emphasis on *linear* error correcting codes. Section 1.5 is a self-contained introduction to information theory, which should provide the reader not familiar with this field with all the basic tools needed to understand this book. In Section 1.6, a short overview of the following chapters is given.

1.2 Modeling a Communication System

Modern point-to-point communication theory is based on a scientific approach to the task of transmitting information from an information source to a remote—either in terms of space or time—destination.

Common to all scientific approaches are the following steps:

- observation of the nature;
- mathematical modeling of the observed phenomena;
- model validation.

Engineering, i.e., designing a communication system, adds a fourth step:

- derivation, on the basis of a mathematical model, of a practical solution to accomplish a given task.

M. Franceschini, G. Ferrari and R. Raheli, *LDPC Coded Modulations,*

DOI: 10.1007/978-3-540-69457-1_1,

The process of modeling is, therefore, a fundamental and central step in the analysis and design of advanced communication systems. In particular, whenever the design target is to obtain a system performance close to the theoretical (ultimate) limits, the quality of the model has a direct impact on the obtainable performance. This can be rigorously quantified with information-theoretic tools [1,2].

In this book, we will focus on the transmission of a digital stream of data, i.e., a discrete-time digital process $\mathcal{A}$ consisting of a sequence of symbols $\{a_k\}$ belonging to a discrete set of possible values $\mathscr{A}$. At the output of the receiver, a corresponding sequence of detected symbols $\hat{\mathcal{A}} = \{\hat{a}_k\}$, $\hat{a}_k \in \mathscr{A}$ is found. The quality of the transmission scheme can be determined by analyzing the difference between the sequence $\mathcal{A}$ and the sequence $\hat{\mathcal{A}}$ and may be expressed through several distinct parameters. Under the assumption of binary symbols, the most used indicator of the transmission quality is the bit error probability, namely, the probability of error for a generic bit in the digital stream, i.e.,

$$P_{\mathrm{e}} \triangleq P\{a_k \neq \hat{a}_k\}.$$

In contexts where the probability $P\{a_k \neq \hat{a}_k\}$ depends on the epoch k, an average error probability can be defined as

$$P_{\mathrm{e}} = \lim_{n \to \infty} \frac{1}{n} \sum_{k=1}^{n} P\{a_k \neq \hat{a}_k\}. \tag{1.1}$$

Depending on the system structure and other assumptions, other quality indicators may be useful, such as the symbol error probability, if the communication system involves the transmission of non-binary symbols, or the frame error probability, if the data stream is partitioned into frames. The error probability is a measure of the rate of occurrence of decision errors over a sufficiently large number of transmission acts according to the interpretation of the concept of probability as the relative frequency of error occurrences. The error rate is a practical performance indicator which is suitable to be measured. In particular the bit error rate (BER), symbol error rate (SER) and frame error rate (FER), represent the rates corresponding to the relevant probabilities. In the following, for brevity, we will not distinguish between error probability and error rate and we will use the two notions interchangably.

Information theory suggests to accomplish reliable transmission by mapping the process $\mathcal{A}$ into a process $\mathcal{C}$, which might be discrete-time or continuous-time, analog or digital, whose statistical properties are suitable for the particular communication channel. This operation is referred to as *channel coding and modulation.* In Figure 1.1, a general point-to-point communication system

Figure 1.1: Point-to-point communication system model.

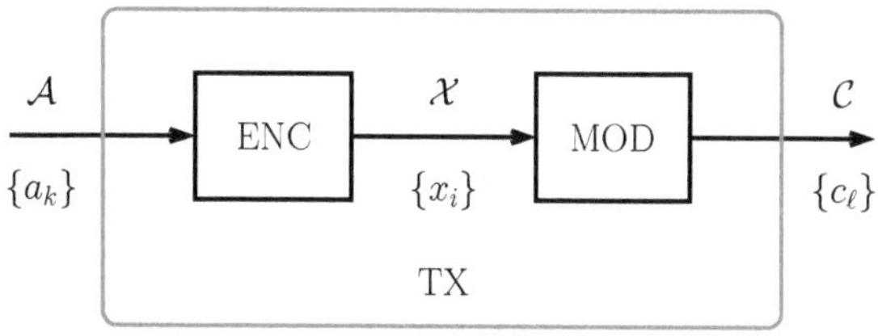

Figure 1.2: Schematic diagram of the transmitter (TX) which performs error correcting coding followed by modulation.

model is shown. Note the presence of the transmitter (TX), whose task is the construction of the signal $\mathcal{C}$ to be transmitted through the channel (CHAN). The receiver (RX) performs detection of the transmitted data sequence based on the received process $\mathcal{R}$. A short introduction to information theory will be given in Section 1.5. For more details, we refer the interested reader to [1,3], which provide a thorough treatment of this subject.

From a practical point of view, in most applications the process $\mathcal{C}$ may be effectively represented by a discrete-time sequence $\{c_\ell\}$. The operation leading from $\mathcal{A}$ to $\mathcal{C}$ is usually split into two separate actions:

- the first consists in performing discrete algebraic operations on $\mathcal{A}$ which result in a new discrete-time digital process $\mathcal{X}$;
- the second consists in mapping the corresponding sequence $\{x_i\}$ into the final sequence $\{c_\ell\}$, to be transmitted through the considered channel.

The first operation corresponds to the application of an error correction encoder to the digital sequence to be transmitted, whereas the second operation, which maps the digital sequence into a signal suitable for transmission over the channel, is usually referred to as modulation. An illustrative scheme of the transmitter performing error correction encoding followed by modulation is shown in Figure 1.2. Note that the sequences $\{a_k\}$, $\{x_i\}$, and $\{c_\ell\}$ have different index variables k, i and ℓ, to highlight the fact that they might not have the same rate and that, on average, a symbol in one of the three sequences does not necessarily correspond to a single symbol in the other sequences.

1.3 Modulation

As stated above, the term *modulation* denotes an operation that maps a sequence of digital symbols, which may correspond to the data to be transmitted or an encoded version of such data, to a sequence of samples or a waveform suitable for transmission through the considered communication channel.

Modulation is a mathematical model of the physical block, found in virtually every communication system, devoted to transform the digital sequence into the transmitted signal, which is often electrical, but may as well be optical, acoustic, etc.

We assume that all the signals in the communication system can be given a discrete-time representation. Modulators are processing blocks and may or may not have memory, meaning that their current output sample may be a function of only the last input sample or more input samples, respectively. Typical memoryless modulations for baseband transmission, which implies mapping the digital sequence into real samples, belonging to a finite set $\mathscr{C}$, referred to as the modulation constellation, include:

- on-off keying (OOK) or, in general, amplitude-shift keying (ASK);
- pulse amplitude modulation (PAM).

A memoryless modulator for bandpass transmission maps the digital sequence into complex samples. Typical complex constellations include:

- quadrature amplitude modulation (QAM);
- phase shift keying (PSK);
- amplitude and phase shift keying (APSK).

For a survey of the principal modulation formats, their constellations, and their properties, the interested reader is referred to [4]. Examples of modulations with memory include:

- differential PSK (DPSK);
- trellis-coded modulation (TCM);
- continuous phase modulation (CPM).

The introduction of memory into the modulator provides potentially infinite degrees of freedom. In fact, any combination of a processing block with a given modulator may be interpreted as a modulator with memory. In order to take advantage of the separation between coding, i.e., the operation leading from $\mathcal{A}$ to $\mathcal{X}$, and modulation, i.e., the operation leading from $\mathcal{X}$ to $\mathcal{C}$, the complexity of the modulation process should be kept low. Nonetheless, it is possible to consider, as part of the tasks carried out by a modulator with memory, operations which usually are considered separately, such as, for example

- pilot symbols insertion;
- a line coding block for spectral shaping.

The separation of coding and modulation is the starting point for the derivation of low-density parity-check (LDPC)-coded modulations. In particular, it allows to separately focus on modulation, whose role is to provide an effective "interface" to the channel, and coding, whose role is to improve the efficiency of information transmission. This will be investigated in depth in Chapter 5.

1.4 Error Correcting Codes

Error correcting codes (ECC) are functions which map a sequence $\mathcal{A}$ of discrete values, i.e., the data sequence, into a (usually longer) sequence $\mathcal{X}$ of discrete values, i.e., the code sequence. As their name suggests, ECC admit inverse functions capable to recover the data sequence $\mathcal{A}$, regardless of possible transmission errors in the code sequence $\mathcal{X}$, provided that the amount of erroneous symbols does not exceed the error correction capability.

Typically, the encoding of a data stream can be performed in two different ways: (i) block coding and (ii) stream coding. The first one consists in partitioning the data sequence into blocks of length K and applying to each block a coding function returning blocks, i.e., the *codewords*, of length N. The obtained codewords are joined to form the code sequence. In stream coding, the data sequence is fed to a finite state machine (FSM), which outputs one or more code symbols for each input data symbol.

1.4.1 Block Codes

In general, a block code C is a vector function associating a vector of K elements, belonging to a set $\mathscr{A}$, to a vector of N elements belonging to a set $\mathscr{X}$. Assuming that the function is injective, the cardinality of the set of all codewords is equal to $[\text{Card}(\mathscr{A})]^K$, where $\text{Card}(\mathscr{A})$ is the cardinality of $\mathscr{A}$. The code rate R_C, expressed in bits per output symbol, is defined as

$$R_C = \frac{K \log_2 \text{Card}(\mathscr{A})}{N} .$$

A linear block code over a q-ary Galois finite field—denoted as GF(q)—is a vector linear function which associates a K-dimensional vector $\boldsymbol{a}$ of elements in GF(q), i.e., the data to be encoded, to a N-dimensional vector $\boldsymbol{x}$ of elements in GF(q), i.e., the codeword, according to the following rule:

$$\boldsymbol{x} = G\boldsymbol{a}$$

where G is a matrix in GF(q) referred to as code generation matrix [5,6]. It can be shown that the set of all codewords can be characterized by the so-called parity check matrix, usually denoted with H, which can be obtained by proper manipulation of G. A vector $\boldsymbol{x}$ is a codeword if and only if

$$H\boldsymbol{x} = 0$$

i.e., the set of all codewords is defined as the null space of the parity check matrix H [5, 6]. In this book, only binary codes will be considered; hence, in all cases data and codeword elements will be in GF(2), i.e., the Boolean algebra.

Several important code families belong to the set of linear block codes, among which:

- cyclic codes, comprising Bose-Chaudhuri-Hocquenghem (BCH) codes, Reed-Solomon codes, Hamming codes [6];
- turbo codes [7,8];
- LDPC codes [9–13].

Several techniques exist for decoding linear block codes. In particular, two approaches are possible: (i) hard-output decoding, which outputs estimates of the codewords or the data symbols, and (ii) soft-output decoding, which outputs the *likelihood* of each codeword or data symbol, i.e., an estimate of how much a codeword or a data symbol is likely to assume each possible value. In this book, we are interested in soft-output decoding only, since, as it will be clear in the next chapters, this leads to the construction of complex receivers based on simpler soft-input soft-output (SISO) detection/decoding blocks.

1.4.2 Stream Codes

Stream coding is an error correcting coding technique which applies to streams of data as opposed to blocks of data. The codeword length is unbounded, so that a stream encoder typically operates on a symbol-by-symbol basis, outputting an encoded symbol for each data symbol at its input. Practical stream encoders are implemented by means of FSMs. In particular, an FSM is defined by two functions: (i) the *output function* and (ii) the *next state function*. These two functions are mathematically described as follows:

$$c_k = f_{\mathrm{o}}(a_k, s_k) \tag{1.2}$$

$$s_{k+1} = f_{\mathrm{ns}}(a_k, s_k) \tag{1.3}$$

where c_k denotes the output symbol at epoch k, s_k denotes the state at epoch k, a_k is the input data symbol at epoch k, and $f_{\mathrm{o}}(\cdot,\cdot)$ and $f_{\mathrm{ns}}(\cdot,\cdot)$ denote the output function and the next state function, respectively. The output stream is formed by the sequence $\{c_k\}$. Without loss of generality, it is assumed that the code sequence has the same length of the information sequence—the redundancy introduced by the encoder is accounted for by expanding the symbol cardinality.

Linear stream coding is the most popular stream coding technique for error correction. Such codes are also known as *convolutional codes* since the output stream can be seen as the convolution of the input data stream with an "impulse response" stream. The convolution is carried out using finite field algebra.

The decoding of stream codes, like that of block codes, can be performed by means of hard-output or soft-output decoding algorithms. The most important hard decoding algorithm for stream coding is the well known Viterbi algorithm (VA) [4, 14, 15]. The VA is optimum, in the sense that it computes the most likely transmitted data sequence given the received observable sequence. Soft decoding/detection techniques for stream codes comprise the soft-output Viterbi algorithm (SOVA) [16].

In general, in stream coding it is not possible to compute exact *a posteriori* probability (APP) of the (possibly infinite) transmitted data symbols. Nevertheless, APP computation is possible, *if the stream encoder is used to encode a finite-length sequence*, by means of the famous Bahl-Cocke-Jelinek-Raviv (BCJR) algorithm, also known as forward backward (FB) algorithm [17].

1.5 Information Theory Basics

In this book, only a basic set of tools drawn from classical information theory will be used. In this section, the main information-theoretic concepts will be concisely presented, to provide the unfamiliar reader with a sufficient (minimum) background. For more details, we refer the interested reader to fundamental information theory textbooks [1,3,18,19] and to the vast scientific literature referenced therein.

While studying communication problems, we are interested in the information transfer capabilities of communication systems. In order to set a mathematical framework, the first problem is to define the concept of information measure. A discrete *information source* $\mathcal{C}$ is an entity emitting a stochastic sequence of symbols $\{C_k\}$, each belonging to a finite set $\mathscr{C}$. Let us assume to describe the sequence as an *ergodic* random sequence. We define the concept of *entropy rate* $\mathcal{H}(\mathcal{C})$ to measure the expected rate of information emitted by the information source $\mathcal{C}$ as follows:

$$\mathcal{H}(\mathcal{C}) = \lim_{n\to\infty} -\frac{1}{n}\mathrm{E}\{\log p(C_1,\ldots,C_n)\} \tag{1.4}$$

where $\mathrm{E}\{\cdot\}$ denotes the expectation, $p(c_1,\ldots,c_n)$ denotes the probability that the first n elements of the random sequence $\{C_k\}$ are equal to $c_1,\ldots,c_n$, and log is the logarithm to a given base. For base 2, information is measured in *bits*. If the source is memoryless, i.e., the output values are independent and identically distributed (i.i.d.), the entropy rate is usually referred to as *entropy* and its definition is the following:

$$\mathcal{H}(\mathcal{C}) = \mathsf{H}(C_k) \triangleq -\mathrm{E}\{\log p(C_k)\} \tag{1.5}$$

which, since $\{C_k\}$ are i.i.d., does not depend on k.

The entropy of a source can be interpreted as a measure of its unpredictability. Sources emitting samples which can be predicted with high probability are characterized by low entropy whereas sources emitting samples which can be predicted with low probability have higher entropy. The entropy rate of a discrete information source is always higher than or equal to zero. In particular, $\mathcal{H}(\mathcal{C}) = 0$ if and only if there exist a sequence $\{c_k^*\}$ such that $P\{C_k = c_k^*\} = 1$, $\forall k$.

In general, any stream operation—deterministic or stochastic—on the output of a source $\mathcal{C}$, leads to a new source $\mathcal{Y}$ with different statistical properties. The *conditional entropy rate* of $\mathcal{Y}$ given $\mathcal{C}$ is defined as follows:

$$\mathcal{H}(\mathcal{Y}|\mathcal{C}) \triangleq \lim_{n\to\infty} -\frac{1}{n}\mathrm{E}\{\log p(Y_1,\ldots,Y_n|C_1,\ldots,C_n)\} \tag{1.6}$$

and measures the "residual" entropy rate of $\mathcal{Y}$ given $\mathcal{C}$.

The *(mutual) information rate* between $\mathcal{C}$ and $\mathcal{Y}$ is defined as follows:

$$\mathsf{I}(\mathcal{C};\mathcal{Y}) = \mathcal{H}(\mathcal{Y}) - \mathcal{H}(\mathcal{Y}|\mathcal{C}) \tag{1.7}$$

and measures the average per-sample amount of information carried by $\mathcal{Y}$ regarding $\mathcal{C}$.

In the case of i.i.d. sources, the conditional entropy rate and the information rate are referred to as *conditional entropy* and *mutual information*, respectively.

Entropy and information rate can be extended to sources emitting continuously distributed samples. In this case, the entropy rate is referred to as *differential entropy rate* and is defined as follows:

$$\mathsf{h}(\mathcal{C}) \triangleq \lim_{n\to\infty} -\frac{1}{n}\mathrm{E}\{\log p(C_1,\ldots,C_n)\} \tag{1.8}$$

where $p(C_1,\ldots,C_n)$ denotes the probability density function (pdf) of the first n elements of the random sequence computed using as argument the (random) vector $(C_1,\ldots,C_n)$. The differential entropy rate may take any real value, i.e., it is not limited to positive or zero values.

If $\mathcal{C}$ is a generic source (with discrete or continuously distributed alphabet), $\mathcal{Y}$ is a source emitting symbols in a continuously distributed alphabet, and $\mathcal{Y}$ and $\mathcal{C}$ are jointly ergodic, the information rate between $\mathcal{C}$ and $\mathcal{Y}$ can be defined in terms of the differential entropy rate as follows:

$$\mathsf{I}(\mathcal{C};\mathcal{Y}) = \mathsf{h}(\mathcal{Y}) - \mathsf{h}(\mathcal{Y}|\mathcal{C}) \tag{1.9}$$

or, equivalently, in terms of the entropy rate as

$$\mathsf{I}(\mathcal{C};\mathcal{Y}) = \mathcal{H}(\mathcal{C}) - \mathcal{H}(\mathcal{C}|\mathcal{Y})\,. \tag{1.10}$$

Unlike the differential entropy rate, the information rate is larger than or equal to zero, regardless of the type of source. As a consequence, if $\mathcal{C}$ in (1.10) is a discrete source, given that

$$\mathsf{I}(\mathcal{C};\mathcal{Y}) = \mathcal{H}(\mathcal{C}) - \mathcal{H}(\mathcal{C}|\mathcal{Y}) \geq 0$$

and, since the entropy rate (of a discrete source) is non-negative,

$$\mathcal{H}(\mathcal{C}|\mathcal{Y}) \geq 0$$

it follows that

$$\mathsf{I}(\mathcal{C};\mathcal{Y}) \leq \mathcal{H}(\mathcal{C})\,.$$

In other words, the information rate between a discrete source and a generic source is upper bounded by the entropy rate of the discrete source.

In [20], C. E. Shannon showed that, if the input and the output of a channel have the same joint statistical properties of two sources $\mathcal{C}$ and $\mathcal{Y}$, respectively, $\mathsf{I}(\mathcal{C};\mathcal{Y})$ is the supremum of the achievable data rates through that channel (considering the input distribution imposed by $\mathcal{C}$).

For a given channel, the corresponding *channel capacity* is a quantity defined as follows:

$$\mathsf{C} = \max_{\mathcal{C}\in\mathcal{K}} \mathsf{I}(\mathcal{C};\mathcal{Y}) \tag{1.11}$$

where the maximization is carried out over all possible input distributions, i.e., input sources, belonging to a set of sources $\mathcal{K}$ satisfying a given constraint or set of constraints. The channel capacity thus represents the supremum of the achievable information rates that can be reliably transmitted through the considered channel.

The following examples will clarify the role of the information theory quantities described above.

Example 1.1 *Binary input AWGN channel.*

Assume to transmit a sequence of i.i.d. binary symbols $\{C_k\}$, where the random variable (RV) C_k may take on values in the set $\mathscr{C} = \{1, -1\}$ and $P\{C_k = 1\} = P\{C_k = -1\} = 1/2$, through a discrete additive white Gaussian noise (AWGN) channel. We will refer to this configuration as *binary input AWGN* (BIAWGN) channel.

The output samples $\{Y_k\}$ can be expressed as:

$$Y_k = C_k + W_k \tag{1.12}$$

where $\{W_k\}$ are zero-mean i.i.d. Gaussian samples with variance σ_w^2. The signal-to-noise ratio (SNR) can be defined as follows:

$$\mathsf{SNR} \triangleq \frac{\mathrm{E}\{|C_k|^2\}}{\mathrm{E}\{|W_k|^2\}} = \frac{1}{\sigma_w^2}\,. \tag{1.13}$$

The entropy of the input source is

$$\mathcal{H}(\mathcal{C}) = \mathsf{H}(C_k) = -\mathrm{E}\{\log P(C_k)\} = -\frac{1}{2}\log\frac{1}{2} - \frac{1}{2}\log\frac{1}{2} = 1 \text{ bit}\,. \tag{1.14}$$

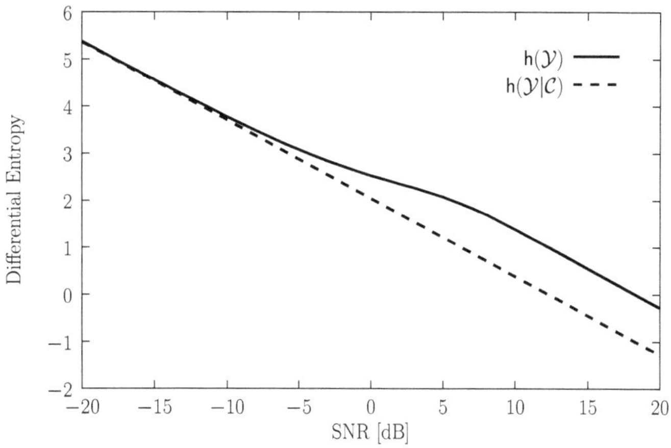

Figure 1.3: Example 1.1: differential entropy rates.

The differential entropy of the output is:

$$
\begin{aligned}
\mathsf{h}(\mathcal{Y}) &= -\mathrm{E}\{\log p(Y_k)\} \\
&= -\int_{-\infty}^{\infty} p(y)\log p(y)\mathrm{d}y \\
&= -\int_{-\infty}^{\infty} \frac{e^{-\frac{(y-1)^2}{2\sigma_w^2}} + e^{-\frac{(y+1)^2}{2\sigma_w^2}}}{2\sqrt{2\pi\sigma_w^2}} \log \frac{e^{-\frac{(y-1)^2}{2\sigma_w^2}} + e^{-\frac{(y+1)^2}{2\sigma_w^2}}}{2\sqrt{2\pi\sigma_w^2}} \mathrm{d}y \qquad (1.15)
\end{aligned}
$$

and the conditional differential entropy of the output given the input is:

$$
\begin{aligned}
\mathsf{h}(\mathcal{Y}|\mathcal{C}) &= \mathsf{h}(\mathcal{W}+\mathcal{C}|\mathcal{C}) \\
&= \mathsf{h}(\mathcal{W}) \\
&= -\mathrm{E}\{\log p(W)\} \\
&= -\int_{-\infty}^{\infty} \frac{1}{\sqrt{2\pi\sigma_w^2}} e^{-\frac{(w)^2}{2\sigma_w^2}} \log \frac{e^{-\frac{(w)^2}{\sigma_w^2}}}{2\sqrt{2\pi\sigma_w^2}} \mathrm{d}w \\
&= \frac{1}{2}\log 2\pi e\sigma_w^2 \,. \qquad (1.16)
\end{aligned}
$$

In Figure 1.3, $\mathsf{h}(\mathcal{Y}|\mathcal{C})$ and $\mathsf{h}(\mathcal{Y})$ are shown as functions of the SNR. The curves are monotonically decreasing, owing to the fact that at higher SNR $\mathcal{Y}$ becomes more predictable. The difference between $\mathsf{h}(\mathcal{Y})$ and $\mathsf{h}(\mathcal{Y}|\mathcal{C})$, however, is a monotonically increasing function of the SNR. In Figure 1.4, the information rate (IR) $\mathsf{I}(\mathcal{Y};\mathcal{C}) = \mathsf{h}(\mathcal{Y}) - \mathsf{h}(\mathcal{Y}|\mathcal{C})$ is shown as a function of the SNR. At very low

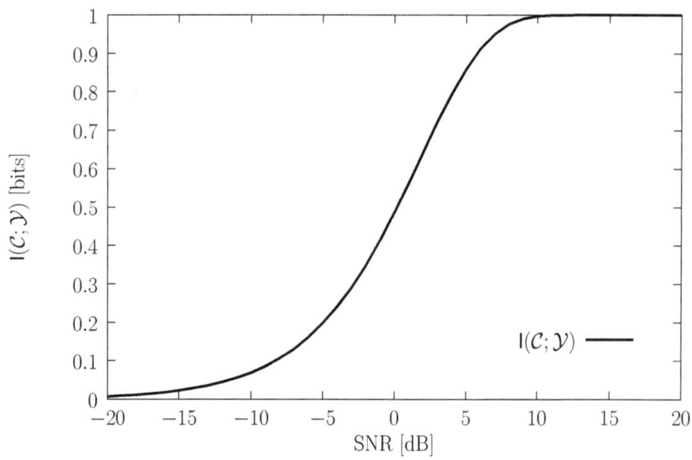

Figure 1.4: Example 1.1: information rate.

SNR, the IR is close to zero and increases as the SNR increases. Its asymptotic value is given by $\mathcal{H}(\mathcal{C}) = 1$. For each SNR value, the IR gives the maximum attainable average information rate (per transmitted binary symbol) assuming the given statistical distribution for $\mathcal{C}$. For example, at 0 dB the IR is equal to 0.486 bit: therefore, the maximum attainable data rate that can be sent through the channel is 0.486 bit per channel use (or bit per sample). In other words, codes with code rate $R_C > 0.486$ shall not achieve arbitrarily low BER at a SNR equal to 0 dB. On the other hand, for every code rate $R_C < 0.486$ there exists a code that will achieve a specified arbitrarily low BER provided that the codeword length is sufficiently large.

Another useful point of view is to interpret the IR theoretical limit in terms of the SNR required to enable achievability of a given code rate R. For example, from the IR curve in Figure 1.4, one can conclude that it is not possible to achieve an arbitrarily small BER with codes whose rate is equal to 0.5 if SNR$<$ 0.18 dB, regardless of the codeword length. This interpretation of the theoretical performance limits suggested by information theory will be used in this book while discussing numerical performance results for LDPC coded modulated schemes.

Example 1.2 *Binary symmetric channel and correction of residual errors.*

Assume to transmit a sequence of i.i.d. binary symbols $\{C_k\}$, where the RV C_k takes a value in the set $\{0, 1\}$ and $P\{C_k = 1\} = P\{C_k = 0\} = 1/2$. The channel outputs a binary symbol $Y_k \in \{0, 1\}$, which may be equal to C_k

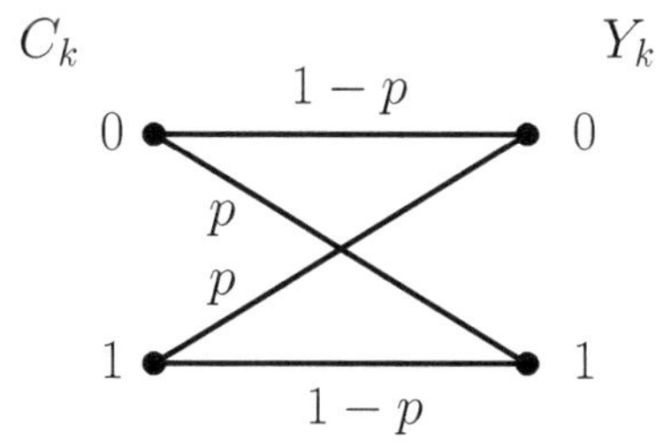

Figure 1.5: Example 1.2: binary symmetric channel.

or $1 - C_k$ according to the following rule:

$$Y_k = \begin{cases} C_k & \text{with probability } 1-p \\ 1 - C_k & \text{with probability } p. \end{cases} \tag{1.17}$$

Hence, p represents the probability of error in a single use of the channel. The channel is memoryless, in the sense that the outputs are conditionally independent, i.e.,

$$P(y_1, \ldots, y_n | c_1, \ldots, c_n) = \prod_{i=1}^{n} P(y_i | c_i) \, .$$

This channel is commonly known as a binary symmetric channel (BSC). In Figure 1.5, a diagram of a BSC is shown. Although this channel is very simple, it may be very useful since almost any digital communication system may be ultimately interpreted as a BSC characterized by p equal to the BER, as defined in (1.1). In fact, if the processes $\mathcal{A}$ and $\hat{\mathcal{A}}$ in Figure 1.1 are binary, they could be seen as the input and the output of a binary-input binary-output channel. To obtain a memoryless binary-input binary-output channel, one may add an interleaving block in front of the input of the transmitter and a corresponding de-interleaving block at the output of the receiver. To make the overall system symmetric, i.e., to obtain equal probability of error for a bit equal to either 0 and to 1, it is sufficient to input the transmitted bit sequence to a "pseudo-random" bit flipper placed before the interleaver and to use a corresponding de-flipper of the bit sequence at the output of the de-interleaver at the receiver side.[1]

[1] Here, *pseudo-random flipping* of a bit sequence denotes flipping of the bits of a subset of the bit sequence, followed, at the receiver, by an identical operation to recover the original bit sequence. For example, one could flip all bits in even positions. In random flipping, each bit is flipped with probability 1/2. This operation is not intended for encryption purposes, but, rather, to shape the statistical characteristics of the bit sequence.

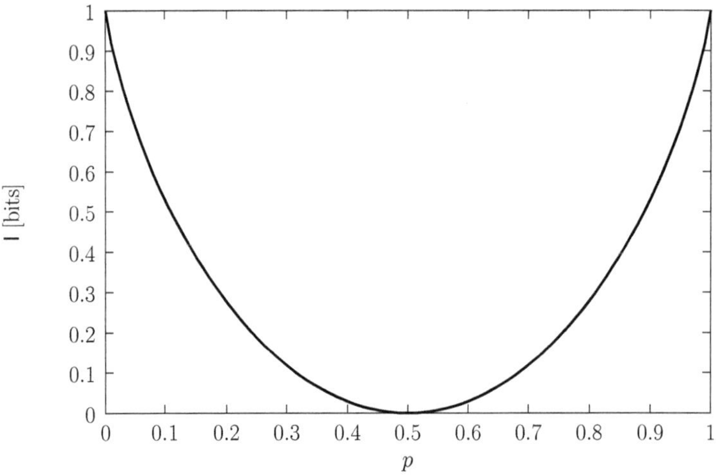

Figure 1.6: Example 1.2: information rate of a binary symmetric channel as a function of the channel error probability p.

The IR of a BSC with the considered input sequence of i.i.d. symbols can be expressed as

$$\begin{aligned} \mathsf{I}(\mathcal{C};\mathcal{Y}) &= \mathcal{H}(\mathcal{Y}) - \mathcal{H}(\mathcal{Y}|\mathcal{C}) \\ &= 1 - H(p) \end{aligned}$$

where

$$H(p) \triangleq -p\log_2 p - (1-p)\log_2(1-p)\,.$$

In Figure 1.6, the IR of a BSC with the considered input sequence of i.i.d. symbols is shown as a function of the BSC transition probability p. By interpreting the IR as the supremum of the achievable code rates, one can obtain interesting insights on systems characterized by a small, but finite, BER.

In particular, consider a communication scheme with a binary process $\mathcal{A}$ at its input and a binary process $\hat{\mathcal{A}}$, consisting of the sequence of decided bits, at its output. The system is assumed to be characterized by a low probability of error $p < 0.5$. By introducing, as discussed earlier, (i) an ideal interleaver before the system input and its corresponding de-interlever at the output of the receiver, and, if necessary, (ii) a bit-flipper and a de-flipper at the transmitter and the receiver, respectively, it is possible to transform the considered communication system into a BSC with transition probability p with arbitrary accuracy. The obtained system is completely characterized by

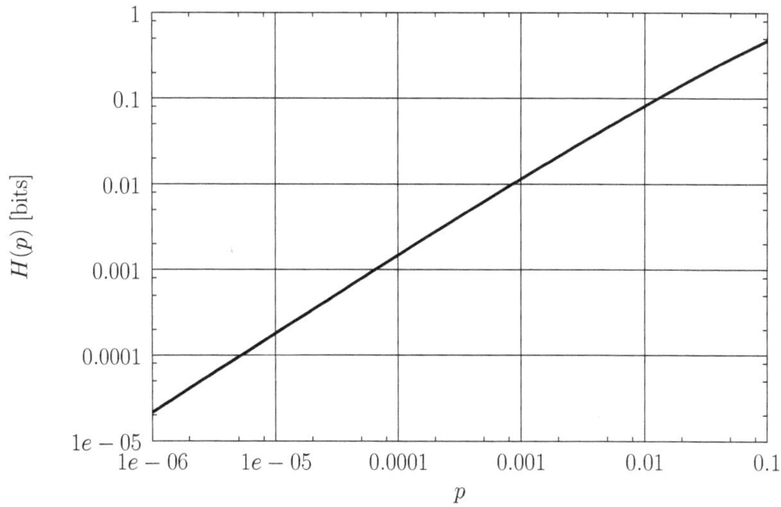

Figure 1.7: Example 2: $H(p)$, i.e., the fraction of bits that must be devoted to outer error correction to obtain arbitrarily small error probability from a communication system characterized by BER equal to p.

its IR, since the function $H(p)$ is invertible for $0 \leq p \leq 0.5$. In particular, by means of a proper outer error correction code for a BSC, with a code rate lower than the IR, it is possible to obtain arbitrarily small BER. In this sense, $1 - \mathsf{I}(\mathcal{C};\mathcal{Y}) = H(p)$ is the minimum fraction of bits that must be "wasted" to guarantee arbitrarily small error probability. In Figure 1.7, the minimum fraction of bits that must be devoted to redundancy for error correction coding of the BSC obtained from a generic communication system is shown considering a BER range from 10^{-6} to 10^{-1}.

For instance, if a communication system achieves a BER equal to 10^{-3}, a fraction at least equal to $H(10^{-3}) \simeq 0.011 \simeq 1\%$ of bits in the transmitted stream must be devoted to guaranteeing an arbitrarily small error rate. In other words, a system with BER equal to 10^{-3} may be used to build a system with arbitrarily low error probability and effective data rate only 1% lower than that of an uncoded system.

1.6 The Following Chapters

This chapter introduced some of the concepts that will be used in the remainder of this book. In particular, basic concepts on the detection techniques of interest will be presented in Chapter 2. The structure and characterization of LDPC codes will be discussed in more detail in Chapter 3. The basic information-theoretic concepts introduced in the current chapter are useful for understanding the theoretical limits that one is facing when designing an advanced communication system. In particular, the extrinsic information transfer (EXIT) chart-based analysis, described in Chapter 4, is based on the concept of mutual information. Bounds, based on mutual information, on the BER of LDPC coded modulated schemes will be given in Chapter 5. Chapter 6 deals with the design of LDPC coded modulations for memoryless channels, whereas Chapter 7 focuses on differentially encoded LDPC coded modulations. Finally, in Chapter 8 some final remarks are provided.

Chapter 2

Trellis-based Detection Techniques

2.1 Introduction

In this chapter, we provide the reader with a brief introduction to the main detection techniques which will be relevant for the low-density parity-check (LDPC) coded modulation schemes considered in the following chapters of this book. In particular, we describe possible trellis-based detection techniques, taking also into account the presence of (binary) coding. Soft-output trellis based detection will be used as component block of LDPC coded modulation receivers in Chapters 5 and 7.

In Section 2.2, we provide a simple classification between hard-output and soft-output detection, presenting the Viterbi algorithm (VA) and the forward-backward (FB) algorithm as key examples of these two detection strategies, respectively. In Section 2.3, we quickly describe optimal detection strategies over channels with memory. In Section 2.4, we consider detection of encoded data. Section 2.5 concludes the chapter.

2.2 Hard-Output and Soft-Output Detection

A general model of a digital transmission system can be viewed as based on both a single M^K-ary signaling act or K repetitions of M-ary signaling acts. In the former interpretation, the message is the entire information sequence, whereas in the latter the message corresponds to an individual information symbol. According to these interpretations, two maximum a posteriori (MAP) detection strategies are obtained. MAP *sequence* detection is optimal in the

M. Franceschini, G. Ferrari and R. Raheli, *LDPC Coded Modulations*,

DOI: 10.1007/978-3-540-69457-1_2,

sense that it minimizes the probability of erroneously detecting the entire sequence, i.e., selecting a sequence not equal to the transmitted one. MAP *symbol* detection minimizes the probability of erroneously detecting each information symbol. More precisely, considering a suitable discretization process, through which the received signal is converted into an equivalent sequence of discrete-time observations $\boldsymbol{r}$, whose dimension depends on the number of samples per symbol interval, the following general formulations of these detection strategies can be derived:

$$\begin{aligned} \hat{\boldsymbol{a}} &= \operatorname*{argmax}_{\boldsymbol{a}} P\{\boldsymbol{a}|\boldsymbol{r}\} && \text{MAP sequence detection} \\ \hat{a}_k &= \operatorname*{argmax}_{a_k} P\{a_k|\boldsymbol{r}\} && \text{MAP symbol detection} \end{aligned}$$

where $\boldsymbol{a} \triangleq \{a_k\}_{k=0}^{K-1}$ and $\boldsymbol{r} \triangleq \{r_k\}_{k=0}^{K-1}$ are the sequences of K information symbols and observables, respectively.[1] As mentioned in the previous section, two algorithms that efficiently implement these two detection strategies are the VA [14,15] and the FB algorithm [17]. Both these algorithms are trellis-based, in the sense that they make use of a *trellis* diagram induced by a finite state machine (FSM) model of the underlying transmitter/channel model. More details will be provided in the remainder of this chapter.

The VA allows to efficiently derive the sequence of *hard* decisions $\hat{\boldsymbol{a}}$. On the opposite, the FB algorithm requires the computation of the a posteriori probability (APP) $P\{a_k|\boldsymbol{r}\}$ of each information symbol. In the case of binary information symbols, a commonly considered reliability value is given by the *logarithmic likelihood ratio* (LLR), derived from the APPs as follows:

$$\text{LLR}_k \triangleq \log \frac{P\{a_k = 0|\boldsymbol{r}\}}{P\{a_k = 1|\boldsymbol{r}\}}. \tag{2.1}$$

It is immediate to recognize that a LLR captures, as a single quantity, the relationship between the APP of a transmitted "1" and that of a transmitted "0." Note that the formulation, based on the use of LLR, can also be extended to the case of larger information symbol cardinality. In the case of M-ary symbols, $(M-1)$ LLRs are needed: the LLR relative to the m-th symbol, $m = 0, \ldots, M-2$, is given by the logarithm of the ratio between $P\{a_k = m|\boldsymbol{r}\}$ and $P\{a_k = M-1|\boldsymbol{r}\}$—in other words, the reference probability is the APP of the last symbol, and its corresponding LLR is thus 0.

[1] Should channel coding or oversampling be used, each information symbol a_k could correspond to more than one observable. However, this can be straightforwardly taken into account by considering a vector notation, i.e., replacing the scalar r_k with a vector $\boldsymbol{r}_k$.

According to the discussion in the previous paragraph, rather than classifying trellis-based detection algorithms depending on the MAP detection strategy (either sequence or symbol), a more practical classification is based on the distinction between hard-output (VA) or soft-output (FB algorithm) detection.

2.2.1 The Viterbi Algorithm

As anticipated at the beginning of this section, the MAP sequence detection strategy can be formulated as

$$\widehat{\boldsymbol{a}} = \operatorname*{argmax}_{\boldsymbol{a}} P\{\boldsymbol{a}|\boldsymbol{r}\}. \tag{2.2}$$

The operation in (2.2) can be described as "finding the sequence $\widehat{\boldsymbol{a}}$ such that the *a posteriori* probability $P\{\widehat{\boldsymbol{a}}|\boldsymbol{r}\}$ of having transmitted $\widehat{\boldsymbol{a}}$, given the received sequence $\boldsymbol{r}$, is maximum." In particular, a conceptual method for the identification of $\widehat{\boldsymbol{a}}$ consists of the evaluation of $P\{\boldsymbol{a}|\boldsymbol{r}\}$ for all possible information sequences $\boldsymbol{a}$, and selection of the sequence which maximizes the a posteriori probability. By chain factorization and owing to the independence of the information symbols, the MAP sequence detection strategy in (2.2) can be formulated in terms of the conditional probability density function (pdf) $p(\boldsymbol{r}|\boldsymbol{a})$ and the a priori sequence probability $P(\boldsymbol{a})$:

$$\begin{aligned} P\{\boldsymbol{a}|\boldsymbol{r}\} \sim p(\boldsymbol{r}|\boldsymbol{a})P(\boldsymbol{a}) &= \prod_{k=0}^{K-1} p(r_k|\boldsymbol{r}_0^{k-1}, \boldsymbol{a})P\{a_k\} \\ &= \prod_{k=0}^{K-1} p(r_k|\boldsymbol{r}_0^{k-1}, \boldsymbol{a}_0^k)P\{a_k\} \\ &\sim \sum_{k=0}^{K-1} \left[\ln p(r_k|\boldsymbol{r}_0^{k-1}, \boldsymbol{a}_0^k) + \ln P\{a_k\}\right] \end{aligned} \tag{2.3}$$

where the symbol $\sim$ indicates that two quantities are monotonically related with respect to the variable of interest (in this case, $\boldsymbol{a}$); a statistical notion of system causality is assumed to hold in the second line; the monotonicity of the logarithm is used in the third line; and $\boldsymbol{r}_0^{k-1}$ is a short-hand notation for the sequence $\{r_i\}_{i=0}^{k-1}$ of k observables. In particular, if the a priori probabilities $\{P\{a_k\}\}$ are equal, the MAP sequence detection criterion coincides with the so-called maximum likelihood (ML) sequence detection criterion.

The maximization of (2.3) over all possible sequences $\{\boldsymbol{a}\}$ can be implemented as a search of a *path* in a *tree* diagram where each branch is in a

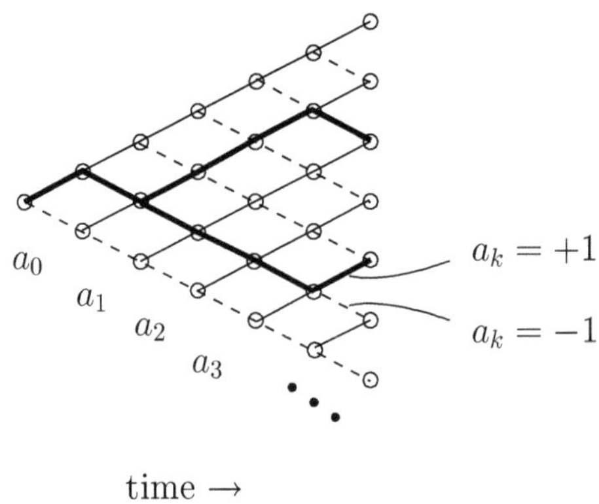

Figure 2.1: Tree representation of the possible transmitted sequences. The number of possible paths to be searched to perform MAP sequence detection increases exponentially with the sequence length.

one-to-one correspondence with an information symbol a_k. Consequently, a path is in a one-to-one correspondence with a *partial* sequence $\boldsymbol{a}_0^k$ up to epoch k. Assigning a *metric* equal to the k-th term of (2.3) to a branch at epoch k associated with symbol a_k and defining a path metric as the sum of the metrics of the branches forming the path, the MAP sequence detection strategy can be implemented as a search of the path with largest metric in this tree diagram. In Figure 2.1, an example of tree comprising all possible paths to be searched is shown, considering a binary alphabet for a_k. Two paths are also emphasized.

While a tree diagram is in principle required, the tree can often be "folded" into a *trellis*, as explained in the following. We assume that the encoder/modulator can be described as a finite state machine (FSM) with state μ_k and characterized by the following "next-state" and "output" functions, respectively:

$$\begin{cases} \mathrm{ns}(\mu_k, a_k) = \mu_{k+1} \\ \mathrm{o}(\mu_k, a_k) = c_k. \end{cases} \tag{2.4}$$

Considering a channel with complex additive white Gaussian noise (AWGN) with circular symmetry, i.e., with independent real and imaginary components, each with variance σ^2—a simple and very common example of memoryless channel—the generic observation at epoch k can be written as

$$r_k = c_k + n_k \tag{2.5}$$

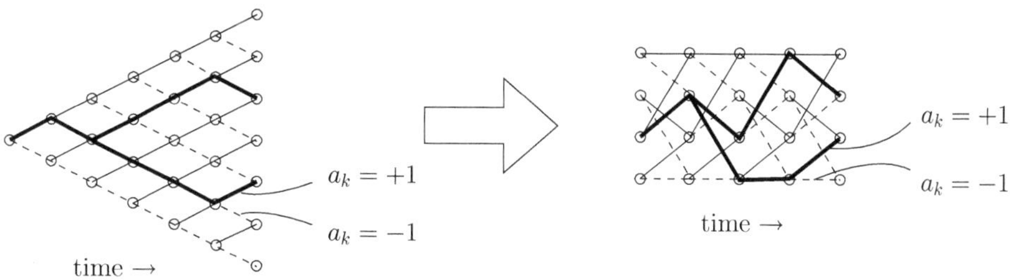

Figure 2.2: Illustrative representation of a tree folded into a trellis.

where n_k is the AWGN sample. In this case, it follows that

$$p(r_k|\boldsymbol{r}_0^{k-1}, \boldsymbol{a}_0^k) = p(r_k|\mu_k, a_k) = \frac{1}{2\pi\sigma^2} e^{-\frac{|r_k - c_k|^2}{2\sigma^2}}$$

and (2.3) can be reformulated as

$$\begin{aligned} P\{\boldsymbol{a}|\boldsymbol{r}\} &\sim \sum_{k=0}^{K-1} \left[-\frac{|r_k - c_k|^2}{2\sigma^2} + \ln P\{a_k\} \right] \\ &\sim \sum_{k=0}^{K-1} \left[-|r_k - c_k|^2 + 2\sigma^2 \ln P\{a_k\} \right]. \end{aligned} \tag{2.6}$$

As mentioned above, a brute-force approach to the implementation of the MAP sequence detection criterion would consist in evaluating (2.6) for all possible information sequences, and choosing the maximizing sequence. Assuming M-ary information symbols, the complexity of this brute-force approach would be M^K, i.e., exponential with the transmission length. Hence, this implementation of the MAP sequence detection principle is feasible only for short transmission lengths, whereas it becomes impractical for transmission lengths of practical relevance in many applications. A much more efficient and appealing MAP sequence detection algorithm is the VA, which will be described in the following.

In the case of strictly finite-memory channels, MAP sequence detection can be formulated as indicated in (2.6), possibly by redefining the symbol c_k and the underlying FSM model. In particular, the optimal tree diagram can be folded into a *trellis diagram*, where the possible states at each epoch are given by all possible values of the state μ_k of the encoder/modulator FSM. In Figure 2.2, an illustrative representation of a folding of a tree into a trellis is shown. In the remainder of this chapter, the number of states of the encoder/modulator FSM will be indicated by S_c. Denoting by $t_k \triangleq (\mu_k, a_k)$ a transition in the trellis diagram, a *branch metric* associated with this transition can be defined as follows:

$$\lambda_k(\mu_k, a_k) = \lambda_k(t_k) \triangleq \ln p(r_k|\mu_k, a_k) + \ln P\{a_k\}. \tag{2.7}$$

Upon the definition of the branch metric (2.7), the a posteriori sequence probability can be written as follows:

$$P\{\boldsymbol{a}|\boldsymbol{r}\} \sim \sum_{k=0}^{K-1} \lambda_k(\mu_k, a_k). \tag{2.8}$$

Without entering into the details (the interested reader can find plenty of literature regarding the VA, e.g. [4, 21]), the implementation principle of the VA is that of associating to each state μ_n a *partial path metric* relative to the corresponding path originating from a known state μ_0, at epoch 0, and terminating into μ_n. This partial path metric, denoted as $\Lambda_n(\mu_n)$, can be written as follows:

$$\Lambda_n(\mu_n) = \sum_{k=0}^{n-1} \lambda_k(t_k) \qquad 1 \le n \le K. \tag{2.9}$$

Obviously, $P\{\boldsymbol{a}|\boldsymbol{r}\} = \Lambda_K(\mu_K)$. Based on the trellis representation of the underlying FSM, the partial metrics, associated with the trellis states, can be computed recursively. For the sake of simplicity, we consider binary information symbols, i.e., $M = 2$. The path metrics associated to states $\mu_k^{(1)}$ and $\mu_k^{(2)}$ are indicated as $\Lambda_k(\mu_k^{(1)})$ and $\Lambda_k(\mu_k^{(2)})$, respectively. The VA associates to state μ_{k+1} (the common ending state of both transitions $t_k^{(1)} = (\mu_k^{(1)}, a_k^{(1)})$ and $t_k^{(2)} = (\mu_k^{(2)}, a_k^{(2)})$) the following path metric:

$$\Lambda_{k+1}(\mu_{k+1}) = \max\left\{\Lambda_k(\mu_k^{(1)}) + \lambda_k(\mu_k^{(1)}, a_k^{(1)}), \Lambda_k(\mu_k^{(2)}) + \lambda_k(\mu_k^{(2)}, a_k^{(2)})\right\}. \tag{2.10}$$

In this sense, the basic operation of the VA is defined as *add-compare-select* (ACS), since: (i) the path metrics associated with the two starting states are summed with the branch metrics of the two branches entering into the common final state; (ii) the obtained partial path metrics are compared; and (iii) the "candidate" largest path metric is selected as *the* path metric associated to μ_{k+1}.

The evaluation of path metrics as indicated above guarantees that the path terminating into any state at a given epoch is, among all entering paths, the one to which the largest metric is associated. At any epoch, the S_c paths with the largest possible path metrics are therefore tracked. Consequently, at the final trellis section at epoch K, the largest path metric among those associated with the final states is such that the corresponding information sequence satisfies the MAP sequence detection criterion in (2.2).

Even if the complexity of the VA is proportional to KS_c (i.e., the dependence on the transmission length is linear and not exponential anymore), the delay would still be unacceptable for large transmission lengths, since one should wait for the transmission of the entire information sequence before being able to output the sequence of symbols satisfying the MAP sequence detection criterion. The VA, however, has an appealing feature which we will describe shortly. At every epoch the number of survivors is equal to the number S_c of states in the trellis. One could track backwards all these survivors starting from the trellis section at epoch k down to the trellis section at epoch $k - D$. For each value of D it is possible to count the number of distinct survivor paths. This number is a monotonically non-increasing function of D. In particular, if the value of D is such that the number of distinct paths is equal to 1, all S_c survivors share the same path from trellis section 0 to trellis section $k - D$. In other words, the survivors merge at section $k - D$. This implies that there is no uncertainty on the best sequence from epoch 0 to epoch $k - D$, and, therefore, the corresponding decisions can be emitted. The probability of having all survivors sharing, at epoch k, the same path in correspondence to trellis section $k - D$ rapidly approaches 1 for increasing values of D. Hence, at epoch k, it is possible to emit the decision $\widehat{a}_{k-D}$ relative to the MAP information sequence which will eventually be selected. In other words, by assuming that consecutive observations are sequentially available, the latency corresponds to only D symbol intervals, where D is sufficiently large to guarantee, at epoch k, high probability of merged survivors at section $k - D$.

2.2.2 The Forward-Backward Algorithm

The most commonly used algorithm to perform MAP symbol detection is the so-called FB algorithm. Seminal work on the design of algorithms for soft-output decoding dates back to the late Sixties [22, 23]. An instance of the FB algorithm was proposed in [24], but a clear formalization is due to Bahl, Cocke, Jelinek and Raviv in 1974 [17]—for this reason, the FB algorithm is also often referred to as BCJR algorithm from the initials of the last names of the authors of [17].

The MAP symbol detection criterion leads to the choice of the symbol $\hat{a}_k$ which minimizes the probability of error with respect to the received signal. More precisely, the MAP symbol detection strategy can be formulated as follows:

$$\hat{a}_k = \operatorname*{argmax}_{a_k} P\{a_k|\boldsymbol{r}\}. \tag{2.11}$$

In order to compute the APP $P\{a_k|\boldsymbol{r}\}$ one can write:

$$P\{a_k|\boldsymbol{r}\} \quad = \quad \sum_{\boldsymbol{a}:a_k} P\{\boldsymbol{a}|\boldsymbol{r}\} \sim \sum_{\boldsymbol{a}:a_k} p(\boldsymbol{r}|\boldsymbol{a})P\{\boldsymbol{a}\} \tag{2.12}$$

where the notation $\boldsymbol{a} : a_k$ denotes all possible information sequences containing a_k, or *compatible* with a_k. Note that the computation of the APP, as indicated in (2.12), can be based on the same metric as in the case of MAP sequence detection (i.e., $p(\boldsymbol{r}|\boldsymbol{a})P\{\boldsymbol{a}\}$) and a further *marginalization* based on the sum over all the information sequences compatible with symbol a_k. Assuming that the information symbols are independent, one can further express the APP as follows:

$$P\{a_k|\boldsymbol{r}\} \quad \sim \quad P\{a_k\} \sum_{\boldsymbol{a}:a_k} p(\boldsymbol{r}|\boldsymbol{a}) \prod_{i=0, i\neq k}^{K-1} P\{a_i\}. \tag{2.13}$$

The first and simplest possible approach for the evaluation of the APP could be based on the computation of expression (2.13). It is immediate to conclude that the computational efficiency of this operation is very low, since one must compute a large number of sequence metrics and then marginalize by adding them together. The complexity would obviously be exponential in the transmission length K. The FB algorithm, introduced in more detail in the following, represents an efficient way to compute the APP, with a complexity linear with the transmission length K, as in the case of a VA for MAP sequence detection.

As already mentioned, the first clear formulation of the FB algorithm can be found in [17]. In the following, we propose a simple derivation of the FB algorithm for transmission over a *memoryless channel*. We assume that the encoder/modulator can be represented as a FSM, with state μ_k and output symbol c_k. We assume that the next-state and the output functions are known[2]

[2] Note that the derivation shown in the following holds also in the case of a channel with strictly finite-memory, the only difference being the interpretation of μ_k as the state

and can be formulated as in (2.4). The couple (μ_k, a_k) uniquely identifies a transition in the trellis diagram of the encoder/modulator FSM. We denote this transition as t_k. After a suitable discretization process with one sample per information symbol, we assume that the observable at epoch k can be written as in (2.5). In this case, the APP can be expressed as follows:

$$\begin{aligned} P\{a_k|\boldsymbol{r}\} &\sim p(\boldsymbol{r}|a_k)P\{a_k\} \\ &= \sum_{\mu_k} p(\boldsymbol{r}|a_k,\mu_k)P\{\mu_k|a_k\}P\{a_k\} \\ &= \sum_{\mu_k} p(\boldsymbol{r}_{k+1}^{K-1}|\boldsymbol{r}_0^k, a_k, \mu_k)p(r_k|\boldsymbol{r}_0^{k-1}, a_k, \mu_k)p(\boldsymbol{r}_0^{k-1}|a_k, \mu_k) \\ &\quad \cdot P\{\mu_k|a_k\}P\{a_k\}. \end{aligned} \tag{2.14}$$

Assuming independent information symbols, since μ_k may depend on $\boldsymbol{a}_0^{k-1}$, it follows that:

$$P\{\mu_k|a_k\} = P\{\mu_k\}.$$

Upon the assumption of transmission over a memoryless channel, the remaining conditional pdfs in (2.14) can be simplified as:

$$\begin{aligned} p(\boldsymbol{r}_{k+1}^{K-1}|\boldsymbol{r}_0^k, a_k, \mu_k) &= p\big(\boldsymbol{r}_{k+1}^{K-1}|\mu_{k+1} = \mathrm{ns}(a_k, \mu_k)\big) \\ p(r_k|\boldsymbol{r}_0^{k-1}, a_k, \mu_k) &= p(r_k|a_k, \mu_k) \\ p(\boldsymbol{r}_0^{k-1}|a_k, \mu_k) &= p(\boldsymbol{r}_0^{k-1}|\mu_k). \end{aligned}$$

Hence, the APP in (2.14) can be rewritten as follows:

$$\begin{aligned} P\{a_k|\boldsymbol{r}\} &\sim \sum_{\mu_k} p\big(\boldsymbol{r}_{k+1}^{K-1}|\mu_{k+1} = \mathrm{ns}(a_k, \mu_k)\big) \\ &\qquad \cdot p(r_k|a_k, \mu_k)p(\boldsymbol{r}_0^{k-1}|\mu_k)P\{\mu_k\}P\{a_k\}. \end{aligned} \tag{2.15}$$

By defining

$$\begin{aligned} \alpha_k(\mu_k) &\triangleq p(\boldsymbol{r}_0^{k-1}|\mu_k)P\{\mu_k\} \\ \gamma_k(a_k, \mu_k) &\triangleq p(r_k|a_k, \mu_k)P\{a_k\} \\ \beta_{k+1}(\mu_{k+1}) &\triangleq p(\boldsymbol{r}_{k+1}^{K-1}|\mu_{k+1}) \end{aligned}$$

---

of the FSM obtained by concatenating the encoder/modulator and the channel. Moreover, we assume generation of a single output symbol c_k in correspondence to each information symbol a_k, but the derivation can be straightforwardly extended to the case of multiple output symbols by using a vector notation, as mentioned in Footnote 1.

the desired symbol APP finally becomes

$$P\{a_k|\boldsymbol{r}\} \sim \sum_{\mu_k} \alpha_k(\mu_k)\gamma_k(a_k,\mu_k)\beta_{k+1}(\mu_{k+1}) \tag{2.16}$$

where, for the sake of notational simplicity, the dependence of μ_{k+1} on μ_k and a_k is not explicitly indicated.[3] In the following, since the generated soft-output value is a quantity monotonically related with the APP, we will indicate this value with the general notation $\mathrm{S}[a_k]$. In other words, one can write:

$$\mathrm{S}[a_k] \triangleq \sum_{\mu_k} \alpha_k(\mu_k)\gamma_k(a_k,\mu_k)\beta_{k+1}(\mu_{k+1}). \tag{2.17}$$

The actual APP values can be obtained by applying a proper normalization to the terms $\mathrm{S}[a_k]$, since $\sum_{a_k} P\{a_k|\boldsymbol{r}\} = 1$. Note that the operation (2.17) where the quantities $\{\alpha_k(\mu_k)\}$ and $\{\beta_{k+1}(\mu_{k+1})\}$ are combined to generate the APP is usually referred to as *completion.*

The quantities $\alpha_k(\mu_k)$ and $\beta_{k+1}(\mu_{k+1})$ can be computed by means of forward and backward recursions, respectively. More precisely, one can write:

$$\begin{aligned}
\alpha_k(\mu_k) &= p(\boldsymbol{r}_0^{k-1}|\mu_k)P\{\mu_k\} \\
&= \sum_{\substack{(\mu_{k-1},a_{k-1}):\\ \mathrm{ns}(a_{k-1},\mu_{k-1})=\mu_k}} p(\boldsymbol{r}_0^{k-1}|a_{k-1},\mu_{k-1},\mu_k)P\{a_{k-1},\mu_{k-1}|\mu_k\}P\{\mu_k\} \\
&= \sum_{\substack{(\mu_{k-1},a_{k-1}):\\ \mathrm{ns}(a_{k-1},\mu_{k-1})=\mu_k}} p(r_{k-1}|\boldsymbol{r}_0^{k-2},a_{k-1},\mu_{k-1},\mu_k) \\
&\qquad \cdot p(\boldsymbol{r}_0^{k-2}|a_{k-1},\mu_{k-1},\mu_k)P\{a_{k-1},\mu_{k-1},\mu_k\} \\
&= \sum_{\substack{(\mu_{k-1},a_{k-1}):\\ \mathrm{ns}(a_{k-1},\mu_{k-1})=\mu_k}} p(r_{k-1}|\boldsymbol{r}_0^{k-2},a_{k-1},\mu_{k-1},\mu_k) \\
&\qquad \cdot p(\boldsymbol{r}_0^{k-2}|a_{k-1},\mu_{k-1},\mu_k)P\{\mu_k|a_{k-1},\mu_{k-1}\} \\
&\qquad \cdot P\{a_{k-1}|\mu_{k-1}\}P\{\mu_{k-1}\}
\end{aligned} \tag{2.18}$$

where the index of the summation indicates all possible transitions $\{(\mu_{k-1}, a_{k-1})\}$ compatible, through the next-state function, with μ_k—this notation is general and accounts also for the case of underlying recursive and non-recursive

[3]This simplifying notational assumption (and other similar assumptions) will also be used in the following. The context should eliminate any ambiguity.

FSM models. The summation in (2.18) can be re-interpreted as carried over all possible trellis branches $\{t_{k-1}\}$ compatible with the final state μ_k. Since we are considering possible combinations of μ_{k-1} and a_{k-1} compatible with μ_k, it follows that

$$P\{\mu_k | a_{k-1}, \mu_{k-1}\} = 1. \tag{2.19}$$

On the basis of the independence between the information symbols and recalling the absence of channel memory, one can write:

$$\begin{aligned} p(r_{k-1} | \boldsymbol{r}_0^{k-2}, a_{k-1}, \mu_{k-1}, \mu_k) &= p(r_{k-1} | \mu_{k-1}, a_{k-1}) \\ p(\boldsymbol{r}_0^{k-2} | a_{k-1}, \mu_{k-1}, \mu_k) &= p(\boldsymbol{r}_0^{k-2} | \mu_{k-1}) \\ P\{a_{k-1} | \mu_{k-1}\} &= P\{a_{k-1}\}. \end{aligned}$$

Finally, a step in the forward recursion in (2.18) can be concisely expressed as follows:

$$\begin{aligned} \alpha_k(\mu_k) &= \sum_{t_{k-1}:\mu_k} p(\boldsymbol{r}_0^{k-2} | \mu_{k-1}) P\{\mu_{k-1}\} p(r_{k-1} | \mu_{k-1}, a_{k-1}) P\{a_{k-1}\} \\ &= \sum_{t_{k-1}:\mu_k} \alpha_{k-1}(\mu_{k-1}) \gamma_{k-1}(t_{k-1}). \end{aligned} \tag{2.20}$$

A similar derivation holds also for the backward recursion. More precisely, one can write:

$$\begin{aligned} \beta_k(\mu_k) &= p(\boldsymbol{r}_k^{K-1} | \mu_k) \\ &= \sum_{a_k} p(\boldsymbol{r}_k^{K-1} | a_k, \mu_k) P\{a_k | \mu_k\} \\ &= \sum_{a_k} p(r_k | \boldsymbol{r}_{k+1}^{K-1}, a_k, \mu_k) p(\boldsymbol{r}_{k-1}^{K-1} | a_k, \mu_k) P\{a_k | \mu_k\}. \end{aligned} \tag{2.21}$$

On the basis of the independence between information symbols and the absence of memory of the considered transmission channel, the following simplifications hold:

$$\begin{aligned} p(r_k | \boldsymbol{r}_{k+1}^{K-1}, a_k, \mu_k) &= p(r_k | a_k, \mu_k) \\ p(\boldsymbol{r}_{k-1}^{K-1} | a_k, \mu_k) &= p(\boldsymbol{r}_{k-1}^{K-1} | \mu_{k+1} = \text{ns}(a_k, \mu_k)) \\ P\{a_k | \mu_k\} &= P\{a_k\}. \end{aligned}$$

A step in the backward recursion, as indicated in (2.21), can be rewritten as follows:

$$\begin{aligned}\beta_k(\mu_k) &= \sum_{a_k} p(\boldsymbol{r}_{k-1}^{K-1}|\mu_{k+1})p(r_k|a_k,\mu_k)P\{a_k\} \\ &= \sum_{a_k} \beta_{k+1}(\mu_{k+1})\gamma_k(t_k).\end{aligned}$$

2.3 Optimal Detection Strategies for Channels with Memory

In this section, we recall the basic detection strategies which can be devised for communication channels with memory. In particular, as seen in the previous sections, the same basic *metric* can be used for both hard-output and soft-output detection. For more details, the reader is referred to [25, 26].

Consider the transmission system model previously described. A sequence of independent and identically distributed M-ary information symbols $\{a_k\}$ are transmitted successively from epoch 0 to epoch $K-1$. The encoder/modulator block can be described as a time-invariant FSM (e.g., a trellis coded modulator, TCM, [27] or a continuous phase modulator, CPM, [28]), and we assume that next-state and output functions can be expressed as in (2.4).

A *causality condition* for the considered communication system can be formulated in terms of statistical dependence of the observation sequence $\boldsymbol{r}_0^k$, up to epoch k, on the information sequence. Accordingly, a system is causal if

$$p(\boldsymbol{r}_0^k|\boldsymbol{a}) = p(\boldsymbol{r}_0^k|\boldsymbol{a}_0^k). \tag{2.22}$$

Similarly, a *finite-memory condition* (FMC) can be formulated, in statistical terms, as follows:

$$p(r_k|\boldsymbol{r}_0^{k-1},\boldsymbol{a}_0^k) = p(r_k|\boldsymbol{r}_0^{k-1},\boldsymbol{a}_{k-C}^k,\mu_{k-C}) \tag{2.23}$$

where C is a suitable *finite-memory parameter* and μ_{k-C} represents the state, at epoch $k-C$, of the encoder/modulator. The considered model includes any definition of state μ_k in terms of a suitable state variable, not necessarily defined in terms of input variables. It can easily be proved that causality and finite-memory conditions imply the following equalities [26]:

$$\begin{aligned}p(r_k|\boldsymbol{r}_0^{k-1},\boldsymbol{a}_{k-D}^k,\mu_{k-D}) &= p(r_k|\boldsymbol{r}_0^{k-1},\boldsymbol{a}_{k-C}^k,\mu_{k-C}) \qquad \forall D \geq C \quad (2.24)\\ p(\boldsymbol{r}_k^{K-1}|\boldsymbol{r}_0^{k-1},\boldsymbol{a}_0^{k-1}) &= p(\boldsymbol{r}_k^{K-1}|\boldsymbol{r}_0^{k-1},\boldsymbol{a}_{k-C}^{k-1},\mu_{k-C}). \quad (2.25)\end{aligned}$$

The first equality formalizes the intuition that considering past information symbols, before epoch $k - C$, adds no further information regarding the observation at epoch k. The second equality formalizes the idea that the finite-memory condition[4] extends to future observations beyond epoch k. Upon the introduction of an output function $o(\cdot, \cdot)$ as in (2.4), causality and finite-memory conditions can be formulated as follows:

$$p(\boldsymbol{r}_0^k | \boldsymbol{c}) = p(\boldsymbol{r}_0^k | \boldsymbol{c}_0^k) \tag{2.26}$$

$$p(r_k | \boldsymbol{r}_0^{k-1}, \boldsymbol{c}_0^k) = p(r_k | \boldsymbol{r}_0^{k-1}, \boldsymbol{c}_{k-C}^k). \tag{2.27}$$

We remark, however, that these conditions involve the transmission channel only and, in general, *do not* imply (2.22) and (2.23). A case of interest may be that of a linear block code followed by a memoryless modulator. In particular, a linear block code is not guaranteed to be causal and finite-memory[5] so that the channel causality (2.26) and finite memory (2.27) do not imply the system causality (2.22) and finite memory condition (2.23).

The introduction of the FMC leads naturally to the definition of augmented trellis state and branch (transition):

$$S_k \triangleq (\boldsymbol{a}_{k-C}^{k-1}, \mu_{k-C})$$

$$T_k \triangleq (S_k, a_k) = (\boldsymbol{a}_{k-C}^{k}, \mu_{k-C}).$$

Then, the following common metric can be used in the VA and FB algorithm:

$$\gamma_k(T_k) \triangleq p(r_k | \boldsymbol{r}_0^{k-1}, \boldsymbol{a}_{k-C}^k, \mu_{k-C}) P\{a_k\}. \tag{2.28}$$

In particular:

- the VA can now be formulated in the logarithmic domain, by defining the branch metric $\lambda_k(T_k) \triangleq \log \gamma_k(T_k)$, and obtaining

$$\log P\{\boldsymbol{a} | \boldsymbol{r}\} \sim \sum_{k=0}^{K-1} \lambda_k(T_k);$$

- the symbol APP computed by the FB algorithm can be finally expressed as

$$P\{a_k | \boldsymbol{r}\} \sim \sum_{S_k} \beta_{k+1}(\mathrm{NS}(S_k, a_k)) \gamma_k(S_k, a_k) \alpha_k(S_k).$$

---

[4]Note that there is a slight difference between the formal definition of the finite memory condition (2.23) and (2.25), since in (2.25) the conditioning information sequence is $\boldsymbol{a}_0^{k-1}$ and does not include symbol a_k. This is, however, expedient for the derivation of the backward recursion of the FB algorithm in Subsection 2.2.2.

[5]*Block-wise* causality and finite-memory must be indeed satisfied.

2.4 Detection of Encoded Data

In this section, we discuss on the applicability of the previously devised strategies in the case of encoded data. The presence of coding makes consecutive transmitted symbols *dependent*. In the case of transmission over channels with memory, it follows that the overall memory is due to the channel *and* the encoder. Depending on how these two memory components are "treated," detection and decoding can be properly combined or separated.

2.4.1 Joint Detection and Decoding

In the case of a channel characterized by parameters affected by stochastic uncertainty, a very general parametric model for the observation r_k is the following:

$$r_k = g(\boldsymbol{a}_{k-L}^k, \mu_{k-L}, \boldsymbol{\xi}_0^k) + w_k \tag{2.29}$$

where L is an integer quantifying the encoding memory (e.g., the memory length of a convolutional encoder), $\boldsymbol{\xi}_0^k$ is a sequence of stochastic parameters independent of $\boldsymbol{a}$, and w_k is an additive noise sample. Under this channel model, the following *conditional Markov property*

$$p(r_k|\boldsymbol{r}_0^{k-1}, \boldsymbol{a}_0^k) = p(r_k|\boldsymbol{r}_{k-N}^{k-1}, \boldsymbol{a}_0^k) \tag{2.30}$$

where N is the order of Markovianity, is sufficient to guarantee a FMC. In fact, (2.30) implies the following [26]:

$$p(r_k|\boldsymbol{r}_0^{k-1}, \boldsymbol{a}_0^k) = p(r_k|\boldsymbol{r}_{k-N}^{k-1}, \boldsymbol{a}_{k-C}^k, \mu_{k-C}) \tag{2.31}$$

where the finite-memory parameter is $C = N+L$. It is immediate to recognize that (2.31) represents a special case of (2.23). As a consequence, all the derivations in the previous section hold by using the "exponential metric[6]" $\gamma_k(T_k) = p(r_k|\boldsymbol{r}_{k-N}^{k-1}, T_k)P\{a_k\}$. In other words, (2.31) is the key relation which "links" the algorithms derived in Section 2.3 with the detection problem over channels with memory. A statistical description of the stochastic channel parameter allows one to compute this exponential metric as [25, 26]

$$\begin{aligned}\gamma_k(T_k) &= \frac{p(\boldsymbol{r}_{k-N}^k|T_k)}{p(\boldsymbol{r}_{k-N}^{k-1}|S_k)}P\{a_k\} \\ &= \frac{\mathrm{E}_{\boldsymbol{\xi}_0^k}\{p(\boldsymbol{r}_{k-N}^k|T_k, \boldsymbol{\xi}_0^k)\}}{\mathrm{E}_{\boldsymbol{\xi}_0^{k-1}}\{p(\boldsymbol{r}_{k-N}^{k-1}|S_k, \boldsymbol{\xi}_0^{k-1})\}}P\{a_k\}.\end{aligned} \tag{2.32}$$

[6]The usual notation in the literature refers to a *metric* in the logarithmic domain. Hence, assuming that $\log \gamma_k$ can be referred to as metric, we refer to γ_k as *exponential metric.*

2.4.2 Separate Detection and Decoding

While in the previous subsection the fundamental metric γ was computed by taking into account simultaneously encoding and channel memories, another possible strategy consists in separating the detection process from the decoding process. This can be carried out, at the receiver side, by considering the concatenation of two blocks:

- the first block, i.e., the detector, computes a posteriori reliability values on the coded symbols, by taking into account the channel memory and exploiting the available statistical channel characterization;

- the second block acts as a "standard" decoder, which receives at its input the reliability values generated by the detector. However, care has to be taken in correctly estimating the distribution of the reliability values generated by the detector (typically, their distribution is well approximated as Gaussian [8]).

In the presence of a stochastic channel with order of Markovianity equal to N, as in Subsection 2.4.1, the detector can make use of the final metric (2.32), the only difference, with respect to Subsection 2.4.1, consisting of the fact that $L = 0$, i.e., $C = N$.

2.4.3 Iterative Detection and Decoding

As briefly anticipated in Chapter 1, the concept of *iterative decoding* was originally introduced by Gallager in his Ph.D. thesis [29] for decoding LDPC codes and was crystallized by Berrou and Glavieux in 1993 with the introduction of turbo codes [7, 8]. In this revolutionary work, the authors showed that a complicated code, with a particular structure, can be decoded efficiently with limited complexity. In particular, they considered a parallel concatenated convolutional code (PCCC), constituted by the parallel concatenation, through interleaving, of two convolutional codes. The receiver is based on two component decoders (corresponding to the two constituent convolutional encoders) which *exchange information* between each other. In Figure 2.3, the basic structure of a turbo decoder, relative to a PCCC, is shown. The two component decoders exchange *soft information* between each other and the interleaver is denoted as Π. More precisely, the decoders exchange a modified version of the APP, the so-called *extrinsic information*, which represents the component of the generated soft output on a symbol not depending on the soft-input information on the same symbol [8]. Referring to the FB algorithm

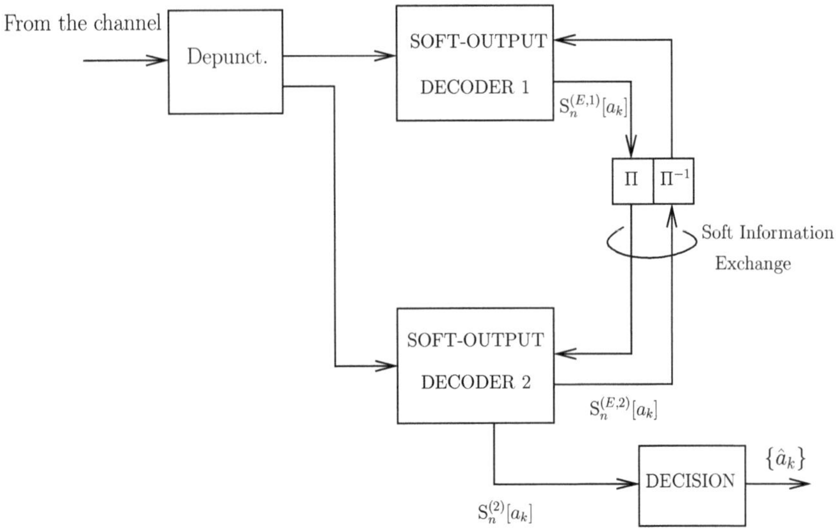

Figure 2.3: Turbo decoder for a PCCC.

formulation proposed in Subsection 2.2.2, the final soft-output quantity, i.e., the extrinsic information, (2.17), can be written as follows:

$$S[a_k] = P\{a_k\}S^{(E)}[a_k]$$

where

$$S^{(E)}[a_k] \triangleq \sum_{\mu_k} \alpha_k(\mu_k)\gamma'_k(a_k, \mu_k)\beta_{k+1}(\mu_{k+1})$$

and

$$\gamma'_k(a_k, \mu_k) \triangleq p(r_k|a_k, \mu_k) = \frac{\gamma_k(a_k, \mu_k)}{P\{a_k\}}.$$

Note that the a priori probability $P\{a_k\}$ of an information symbol used by each component decoder is given, in the iterative decoding process, by the extrinsic information generated by the other decoder, i.e.,

$$P\{a_k\} \longleftarrow S^{(E,i)}[a_k] = \frac{\sum_{\mu_k} \alpha_k(\mu_k)\gamma_k(a_k, \mu_k)\beta_{k+1}(\mu_{k+1})}{S^{(E,j)}[a_k]} \tag{2.33}$$

where $i, j \in \{1, 2\}$, $i \neq j$, and the operator "$\longleftarrow$" denotes a value assignment.

In Figure 2.3, the extrinsic information values on information symbol a_k generated by the first and second decoders at the n-th iteration are denoted by $\mathrm{S}_n^{(E,1)}[a_k]$ and $\mathrm{S}_n^{(E,2)}[a_k]$, respectively. Assuming, as a convention, that an iteration is constituted by the sequence of decoding acts of the first and second component decoders, the soft-output values generated by each component decoder can be re-written as follows:

$$\underbrace{\mathrm{S}_n^{(1)}[a_k]}_{\text{complete output}} = \underbrace{\mathrm{S}_{n-1}^{(E,2)}[a_k]}_{\text{input}} \underbrace{\mathrm{S}_n^{(E,1)}[a_k]}_{\text{extrinsic output}} \tag{2.34}$$

$$\mathrm{S}_n^{(2)}[a_k] = \mathrm{S}_n^{(E,1)}[a_k]\, \mathrm{S}_n^{(E,2)}[a_k]. \tag{2.35}$$

In other words, the soft-output value generated by the first decoder at the n-th iteration is the product between the soft value at its input, corresponding to the extrinsic information generated by the second decoder at the $(n-1)$-th iteration, and the generated extrinsic information. The soft-output value generated by the second decoder at the n-th iteration is the product between the soft value at its input, corresponding to the extrinsic information generated by the first decoder at the same iteration, and the generated extrinsic information. The two soft-output values in (2.34) and (2.35) indicate clearly that the soft-output decoding processes (based on the FB algorithm) in the two component decoders are *coupled*. If the iterative decoding process starts with decoder 1, then the soft-output information (2.35) produced by decoder 2 will be the final soft information at the output.

In the presence of channels with memory, this iterative decoding scheme can be straightforwardly extended to an iterative joint detection/decoding scheme, the only difference being the fact that each component decoder in Figure 2.3 makes use of an FB algorithm based on the metric developed in Subsection 2.4.1.

2.4.4 Turbo Detection

In the previous subsection we have considered iterative joint detection/decoding, where every component block was aware of both channel and code structure. A separate detection/decoding approach, obtained through a serial concatenation of a detector and decoder (as described in Subsection 2.4.2), leads to an iterative receiver scheme where the detector (or soft demodulator), which accounts for the statistical characterization of the channel, and the decoder, which accounts for the code structure, exchange soft information. More precisely, in the non-iterative separate scheme introduced in Subsection 2.4.2, the inner detector computes and passes soft-output information to the outer decoder without a feedback from the decoder itself. The receiver becomes iterative as soon as the outer decoder feeds soft information back to the detector. In the literature, this iterative separate detection and decoding scheme is usually referred to as *turbo-equalization* [30], despite this terminology is slightly abused because, strictly speaking, an equalization process does not take place. An alternative terminology is "turbo detection."

2.5 Concluding Remarks

In this chapter, we have summarized basic detection theoretic concepts which will be used in the following chapters. In particular, we have discussed a unified approach to trellis-based detection over channels with memory. The concepts of joint and separate detection/decoding have been presented, together with a short discussion on iterative detection and turbo equalization. The FB algorithm—or other trellis-based soft-output detection algorithms derived from it, such as, for example, the multi-trellis soft-input soft-output (SISO) algorithm which will be presented in Section 7.6—will be used in the remainder of this book as component blocks to build LDPC coded modulation receivers.

Chapter 3

Low-Density Parity-Check Codes

3.1 Introduction

In this chapter, we provide the reader with an overview on low-density parity-check (LDPC) codes. The concepts outlined in this chapter will then be used througout the remainder of the book.

This chapter is structured as follows. In Section 3.2 a basic description of LPDC codes, together with their graphical representation, is given. In Section 3.3, LDPC codes are described through a statistical approach, which allows to derive significant insights into the behavior of these codes in a tractable manner. In Section 3.4, possible decoding algorithms are presented. Section 3.5 presents LDPC code design techniques based on the use of the statistical description introduced in Section 3.3. Finally, encoding techniques are presented in Section 3.6 and conclusions are drawn in Section 3.7.

3.2 Description of LDPC Codes

LDPC codes were first introduced by R. Gallager in his Ph.D. thesis [9]. In their first instance, LDPC codes are linear block codes characterized by a *sparse* parity-check matrix H whose columns have a fixed number d_v of non-zero elements and whose rows have a fixed number d_c of non-zero elements. The following matrix gives an example of a possible LDPC code parity check

M. Franceschini, G. Ferrari and R. Raheli, *LDPC Coded Modulations*,

DOI: 10.1007/978-3-540-69457-1_3, © Springer-Verlag Berlin Heidelberg 2009

matrix with $d_v = 3$ and $d_c = 6$:

$$\begin{pmatrix} 0&1&0&1&1&0&1&0&0&1&0&0&0&0&0&0&0&0&0&1 \\ 1&0&0&0&0&0&0&1&0&0&0&0&0&1&0&1&1&0&0&1 \\ 0&0&1&0&0&1&0&0&0&0&0&0&1&1&1&1&0&0&0&0 \\ 0&0&1&1&0&0&0&0&0&0&1&0&1&0&0&0&1&1&0&0 \\ 0&1&0&0&1&0&0&1&0&0&0&1&0&0&0&1&0&1&0&0 \\ 0&1&0&0&0&1&1&0&0&0&0&0&0&0&1&0&0&1&0&1 \\ 1&0&0&1&0&0&0&0&1&1&1&0&0&0&0&0&0&0&1&0 \\ 0&0&1&0&0&1&1&0&1&1&0&1&0&0&0&0&0&0&0&0 \\ 1&0&0&0&0&0&0&1&0&0&1&0&0&1&1&0&0&0&1&0 \\ 0&0&0&0&1&0&0&0&1&0&0&1&1&0&0&0&1&0&1&0 \end{pmatrix}. \tag{3.1}$$

In the current literature such codes are referred to as *regular* (d_v, d_c) LDPC codes. Let N be the number of columns of the parity check matrix, i.e., N is the length of the codeword. Let $M < N$ be the number of rows of the parity check matrix and assume that the parity check matrix H is maximum rank, i.e., $\text{Rank}(H) = M$. Considering a column vector $\boldsymbol{x}$ of N elements, the number of degrees of freedom in the solution of the parity check equation

$$H\boldsymbol{x} = 0 \tag{3.2}$$

i.e., the number of linearly independent columns of H, is equal to $K = N - M$, which also corresponds to the number of information bits in a codeword. Of course, the sum of all ones in the rows equals the sum of all ones in the columns. Therefore

$$Nd_v = Md_c = (N - K)d_c$$

which yields

$$\frac{K}{N} = 1 - \frac{d_v}{d_c}. \tag{3.3}$$

The term $R = K/N$ in (3.3) is the so called *code rate*, i.e. the average number of information bits per codeword binary symbol. Therefore, a regular (d_v, d_c) LDPC code has code rate $R = 1 - d_v/d_c$.

It is possible to associate H with a bipartite graph in one-to-one correspondence. Such a graph contains two kinds of nodes: each node of the first kind, denoted as *variable node*, is associated with a column of H; each node of the second kind, denoted as *check node*, is associated with a row of H. The bipartite graph associated with the parity check matrix is constructed as follows:

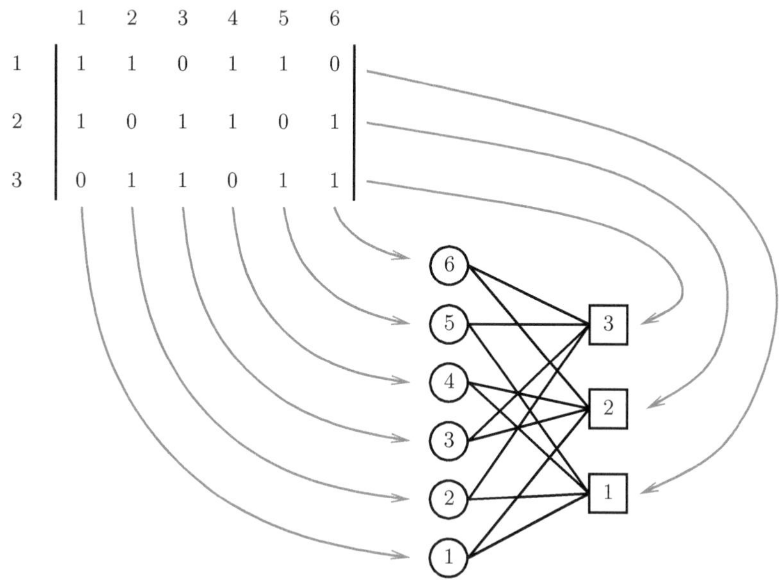

Figure 3.1: Pictorial exemplification of the graph construction for a regular (2, 4) LDPC code.

1. allocate an array of N variable nodes, each one in correspondence to a column of H;

2. allocate an array of M check nodes, each one in correspondence to a row of H;

3. for each nonzero entry of H, connect with a branch the variable and check nodes corresponding to the entry column and row, respectively.

In Figure 3.1, a pictorial exemplification of the construction of the bipartite graph for a very simple LDPC code is shown, characterized by $d_v = 2$ and $d_c = 4$.

In the parity check equation (3.2), each column of the H matrix is multiplied by a corresponding binary symbol in the codeword. Therefore, each variable node is associated with a binary symbol in the codeword. The bit position is represented by the position of the column associated with the variable node. On the other hand, each check node is associated with a parity check equation specified by the row corresponding to that particular check node. A node is said to have "degree i" if i branches depart from it and connect to i different nodes of the other kind.

The above described bipartite graphs for LDPC codes are instances of the so called Tanner graphs for linear block codes. Tanner graphs were introduced by M. Tanner in [31], and may be used to describe the constraints that a codeword must fulfill in order to belong to a particular linear block code. One may observe that, substituting any row in the parity check matrix H with a linear combination of the row itself and any other set of rows does not alter the set of codewords fulfilling the parity check equation (3.2), i.e., the obtained matrix is a parity check matrix for the same code. In other words, every linear block code admits several parity check matrices, or, equivalently, several Tanner graph representations. Nevertheless, in general, the above described linear combination method does not preserve sparseness of the parity check matrix.

It is interesting to note that, although it is possible to construct a parity check matrix for a regular (d_v, d_c) code when d_v is an *even* number, this matrix will not have maximum rank, since there exists at least one linear combination of rows equal to the all-0 vector. In fact, consider the vector $\boldsymbol{r} = (r_1, \ldots, r_N)$ obtained by summing all row vectors in H. The j-th element of $\boldsymbol{r}$ is

$$\begin{aligned} r_j &= \sum_{i=1}^{M} h_{ij} \mod 2 \\ &= d_v \mod 2 \\ &= 0 \end{aligned}$$

where $\{h_{ij}\}$ denote the entries of H. Therefore, it is possible to remove a row from the parity check matrix without modifying the set of codewords and increasing the code rate. As a consequence, there cannot exist regular (d_v, d_c) LDPC codes with even d_v and rate $1 - d_v/d_c$.

3.3 Statistical Description of LDPC Codes

The notation (d_v, d_c) for regular LDPC codes describes a peculiar property, i.e., that the code admits at least a parity check matrix that has exactly d_v ones in each column and d_c ones in each row. There can be more than one actual code that has this property, even if the codeword length N is fixed. Therefore, this notation identifies a class or *ensemble* of codes. Several LDPC codes performance analysis techniques refer to code ensembles, in the sense that the analysis characterizes the expected performance when an actual code is randomly selected within an ensemble.

Although the regularity assumption greatly simplifies performance analysis, it imposes unnecessary constraints on the structure of the parity check matrix. As a consequence, in [10,13,32] *irregular* LDPC codes have been proposed, which allow for a different number of non-zero elements in each row and column.

The bipartite graph construction does not rely on code regularity and therefore applies to irregular LDPC codes as well. In the irregular case, the number of branches connected to the various nodes may change from node to node, regardless the node kind. In other words, variable (or check) nodes are not constrained to have equal degree.

To give a description of irregular LDPC codes ensembles, in [10, 13, 32], the concept of *degree distribution* is introduced.[1] The degree distributions of an LDPC code is specified by a pair of polynomials

$$(\lambda(x), \rho(x)) \triangleq \left(\sum_{i=1}^{\infty} \lambda_i x^{i-1}, \sum_{j=1}^{\infty} \rho_j x^{j-1} \right)$$

where the coefficient λ_i is the fraction of graph branches connected to degree-i *variable* nodes and ρ_j is the fraction of graph branches connected degree-j *check* nodes [13]. The polynomial $\rho(x)$ is referred to as the *check node degree distribution* and $\lambda(x)$ is referred to as *variable node degree distribution.* The coefficients $\{\rho_j\}$ and $\{\lambda_i\}$ must satisfy the following constraints [13]:

$$\begin{array}{rcll} 0 \leq & \rho_j & \leq 1 & j \geq 1 \\ 0 \leq & \lambda_i & \leq 1 & i \geq 1 \\ \sum_{j=1}^{\infty} \rho_j & = & 1 & \\ \sum_{i=1}^{\infty} \lambda_i & = & 1 & \end{array} \tag{3.4}$$

where the third and the fourth relation arise because the "sum of all fractions of edges" must be equal to one.

If a graph has l branches, i.e., the corresponding parity check matrix has l nonzero entries, the number v_i of degree-i variable nodes is

$$v_i = \frac{l\lambda_i}{i}$$

and the number c_j of degree-j check nodes is

$$c_j = \frac{l\rho_j}{j}.$$

[1] Note that the concept of degree distribution is borrowed from random graph theory, where it is characterized by a slightly different definition.

Therefore the number N of variable nodes is given by

$$N = \sum_i v_i = l \sum_i \frac{\lambda_i}{i} = l \int_0^1 \lambda(x)\,\mathrm{d}x$$

where the integral notation is sometimes used in the literature for conciseness. The number M of check nodes is given by

$$M = \sum_j c_j = l \sum_j \frac{\rho_j}{j} = l \int_0^1 \rho(x)\,\mathrm{d}x\,.$$

The code rate $R = K/N$, therefore, is as follows:

$$\begin{aligned} R &= \frac{K}{N} = \frac{N-M}{N} = 1 - \frac{M}{N} \\ &= 1 - \frac{l \sum_j \frac{\rho_j}{j}}{l \sum_i \frac{\lambda_i}{i}} = 1 - \frac{\int_0^1 \rho(x)\,\mathrm{d}x}{\int_0^1 \lambda(x)\,\mathrm{d}x} \end{aligned}$$

One can observe that, given the code rate R, the degree distribution coefficients must satisfy the following linear constraint:

$$\sum_{j=1}^{\infty} \frac{\rho_j}{j} = (1-R) \sum_{i=1}^{\infty} \frac{\lambda_i}{i}\,. \tag{3.5}$$

Irregular LDPC codes are among the most powerful binary codes known today. In [13], it is shown how to design the degree distributions of powerful irregular LDPC codes and in [33] it is shown that carefully designed irregular LDPC codes can practically achieve performance as close as 0.0045 dB to the Shannon limit of the additive white Gaussian noise (AWGN) channel. Irregular LDPC codes characterized by performance close to the capacity limit have been obtained for a variety of binary input memoryless channels, among which the binary erasure channel (BEC) and the binary symmetric channel (BSC).

3.4 Decoding Algorithms for LDPC Codes

3.4.1 Sum-Product Algorithm

Maximum a posteriori probability (MAP) decoding, either per-symbol or per-codeword, of a generic linear block code is, in general, a formidable task.

However, in [9] Gallager introduces three suboptimal iterative decoding algorithms for LDPC codes, which exploit the sparseness of the parity check matrix of the code. Two of them have very low complexity and are based on hard decisions and a bit-flipping technique, whereas the third one is a more accurate algorithm which is based on the iterative exchange of real-valued reliabilities of the codeword bits. These algorithms are widely known as Gallager A,B, and C algorithms, after [34].

All these algorithms have the appealing property of being based on the Tanner graph representation of the LDPC code. In particular, they are characterized by the fact that the nodes, both variable and check, act as processors exchanging real-valued messages on the code graph. All the processing is done *locally*, i.e., for each node it is based only on the available messages. The messages represent reliability values for the codeword bits; in particular they represent an estimate of the probability that each particular codeword bit is equal to "1". In [12, 35], the author presents a reinvention of LDPC codes, and the relevant iterative decoding algorithm. He also highlights how the iterative decoding algorithm can be seen as a particular instance of the *belief propagation* (BP) algorithm [36]. In [37], the authors present a general graph-based algorithm, i.e., the *sum-product* (SP) algorithm, which can be useful for computing the marginalization of complex probability density functions. The authors show how BP can be seen as an instance of the SP algorithm and that the SP algorithm achieves optimality if the code graph has no *cycles*. In other words, the Gallager C decoding algorithms computes the exact *a posteriori* probability of each codeword symbol—thus enabling MAP symbol detection—if the code graph has the shape of a tree. This fact had been argued in [9] as well.

The Gallager C algorithm is now introduced with emphasis on its implementation in the logarithmic domain. Given a binary random variable X, taking values in the set $\{0, 1\}$, its likelihood ratio λ_X is defined as

$$\lambda_X = \frac{P\{X = 0\}}{P\{X = 1\}} \tag{3.6}$$

and the corresponding log-likelihood ratio is

$$\Lambda_X = \log \lambda_X \,.$$

At each iteration in the decoding algorithm: (i) first, each variable node computes an output message for each connected edge and, (ii) then, each check node computes an output message for each connected edge. The order of computation of the messages is usually referred to as *schedule*. This is only the

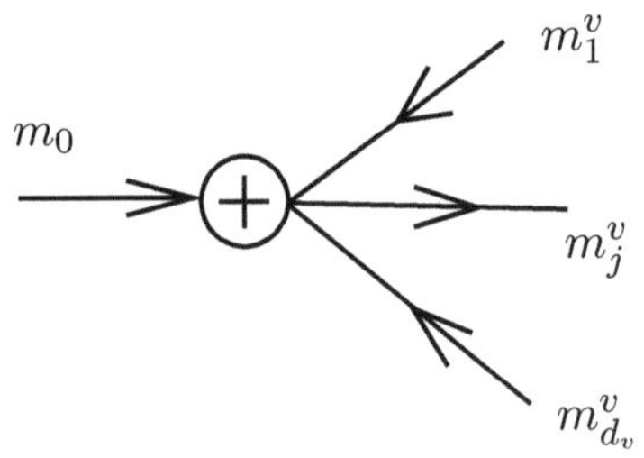

Figure 3.2: Variable node: the quantities involved in the computation of the j-th output message are shown.

most commonly adopted schedule; other schedules exist and can be found in the literature [38–40].

In Figure 3.2, a degree-d_v variable node is shown and the input and output messages are explicitly indicated. One can observe that there is an input, connected to the node and whose value is labeled with m_0, which represents an input reliability value associated with the bit corresponding to the variable node, expressed in the log-likelihood domain. This input reliability value is usually computed on the basis of the observation of the channel.

The decoding algorithm can be derived by reasoning in the probability domain and then by casting the result in the log-likelihood ratio (LLR) domain as follows. The message at the output of a variable node is the probability that the corresponding codeword bit X is equal to "1" given a set of independent observations regarding the bit. Assume to have d_v independent observations $\xi_1, \ldots, \xi_{d_v}$. Let the likelihood ratios of the probability of the observations be

$$\left(\frac{p(\xi_1|X=0)}{p(\xi_1|X=1)}, \ldots, \frac{p(\xi_{d_v}|X=0)}{p(\xi_{d_v}|X=1)}\right) = (\lambda_1, \ldots, \lambda_{d_v}).$$

The probability of X being equal to 0 given the observations is

$$P(X=0|\xi_1, \ldots, \xi_{d_v}) = \frac{p(\xi_1, \ldots, \xi_{d_v}|X=0)P(X=0)}{p(\xi_1, \ldots, \xi_{d_v})}$$

and the corresponding likelihood ratio is

$$\begin{aligned}\lambda &= \frac{P(X=0|\xi_1,\ldots,\xi_{d_v})}{P(X=1|\xi_1,\ldots,\xi_{d_v})}\\ &= \frac{p(\xi_1,\ldots,\xi_{d_v}|X=0)P(X=0)}{p(\xi_1,\ldots,\xi_{d_v}|X=1)P(X=1)}\\ &= \frac{\prod_{i=1}^{d_v} p(\xi_i|X=0)}{\prod_{i=1}^{d_v} p(\xi_i|X=1)}\frac{P(X=0)}{P(X=1)}\\ &= \frac{P(X=0)}{P(X=1)}\prod_{i=1}^{d_v}\lambda_i \qquad (3.7)\end{aligned}$$

which, assuming $P(X=0)=P(X=1)=1/2$, in the LLR domain becomes[2]

$$\Lambda = \sum_{i=1}^{d_v}\Lambda_i \qquad (3.8)$$

where $\Lambda_i = \log\lambda_i$.

The variable node decoding algorithm can be formulated in the logarithmic likelihood domain as follows [9]. Each degree-d_v variable node, as shown in Figure 3.2, computes d_v output messages as follows:

$$m_j^v = m_0 + \sum_{\substack{i=1\\ i\neq j}}^{d_v} m_i^v \qquad (3.9)$$

where m_j^v is the j-th output message and m_i^v is the i-th input message coming from a check node. In other words, the variable node treats all the messages at its input as independent observations: $d_v - 1$ are from the check nodes and one, m_0, corresponds to the likelihood ratio associated with an external observation. This external observation can be a sample from the channel, or it might correspond to an *a priori* information on the corresponding bit. The message m_0 enables to account for an a priori information and will be used in Chapter 5 as a feedback input in a more general decoding scheme.

One can observe that the sum explicitly excludes the message coming from the edge whose output message is being computed. This is in agreement with

[2]We remark that the *a priori* probability $P(X=0)$, and in particular its corresponding likelihood ratio $P(X=0)/P(X=1)$ plays the same role in (3.7) as any other likelihood ratio observation. In other words, it could be formally treated as an observation and embedded into the product.

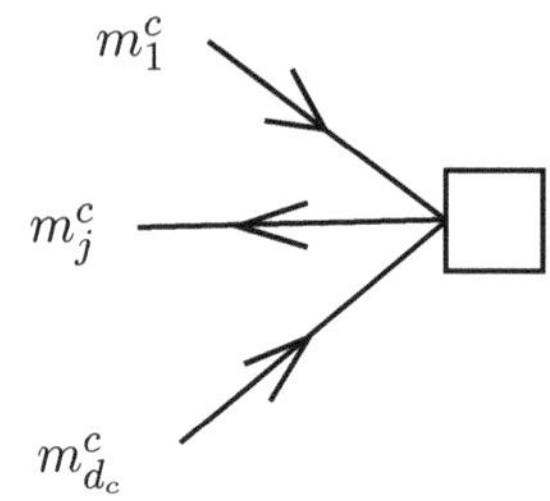

Figure 3.3: Check node: the quantities involved in the computation of the j-th output message are shown.

the use of the so-called *extrinsic information* in iterative detection, and, in particular, in turbo decoding [8].

In Figure 3.3, a generic degree-d_c check node is shown. A check node represents a constraint on the codeword bits associated to the variable nodes connected to it. This constraint is expressed by the corresponding parity check matrix row: *the modulo-2 sum of the codeword bits connected to the check node is equal to 0.* In this case, the associated problem is the following. Given probabilities

$$(P(X_1 = 1), \ldots, P(X_{d_c-1} = 1)) = (p_1, \ldots, p_{d_c-1})$$

associated with $d_c - 1$ bits of the d_c bits connected to the parity check node, compute the probability P_0 that their sum modulo-2 is equal to 0 (we know that the sum of *all* d_c bits modulo-2 is equal to 0). Assuming that all the observations leading to the computation of $P(X_1 = 1), \ldots, P(X_{d_c-1} = 1)$ are independent, then

$$\begin{aligned} P_0 &= \sum_{\boldsymbol{x}:\text{sum is even}} P(\boldsymbol{x}) \\ &= \sum_{\boldsymbol{x}:\text{sum is even}} \prod_{j=1}^{d_c-1} P(x_j) \end{aligned}$$

where "$\boldsymbol{x}$: sum is even" denotes all bit vectors $\boldsymbol{x}$ of length $d_c - 1$ whose sum is an even number, i.e., $\boldsymbol{x}$ contains an even number of 1's. The computation of this quantity can be performed following the guidelines in [9]. First, consider the following polynomial in t

$$q(t) = \alpha_0 + \alpha_1 t + \ldots + \alpha_{d_c-1} t^{d_c-1} = \prod_{j=1}^{d_c-1} (1 - p_j + p_j t) .$$

Observe that the coefficient α_i is given by

$$\alpha_i = \sum_{k_1<k_2<\ldots<k_i} p_{k_1}\ldots p_{k_i} \prod_{j\neq k_1,\ldots,k_i} (1-p_j)\,.$$

In other words, α_i is the probability of having i ones and d_c-1-i zeros among the d_c-1 bits. Observe now that $q(1)$ is the sum of all coefficients and $q(-1)$ is the sum of all coefficients where all odd coefficients have changed sign. Therefore

$$q(1)+q(-1) = \sum_{j=0}^{d_c-1} (1+(-1)^j)\alpha_j = \sum_{k=0}^{\lfloor (d_c-1)/2 \rfloor} 2\alpha_{2k}$$

is equal to two times the sum of all even coefficients, i.e., twice the probability of having an even number of 1's. As a consequence,

$$P_0 = \frac{q(1)+q(-1)}{2} = \frac{1+\prod_{j=1}^{d_c-1} 1-2p_j}{2}.$$

The corresponding LLR is

$$\begin{aligned}
\Lambda &= \log\frac{P_0}{1-P_0} \\
&= \log\frac{1+\prod_{j=1}^{d_c-1}(1-2p_j)}{1-\prod_{j=1}^{d_c-1} 1-2p_j} \\
&= \log\frac{1+\prod_{j=1}^{d_c-1}(1-\frac{2}{e^{\Lambda_j}+1})}{1-\prod_{j=1}^{d_c-1} 1-\frac{2}{e^{\Lambda_j}+1}} \\
&= \log\frac{1+\prod_{j=1}^{d_c-1}\tanh(\Lambda_j/2)}{1-\prod_{j=1}^{d_c-1}\tanh(\Lambda_j/2)} \\
&= 2\text{atanh}\prod_{j=1}^{d_c-1}\tanh(\Lambda_j/2)
\end{aligned}$$

where

$$\Lambda_j = \log\frac{1-p_j}{p_j}\,.$$

Recalling the check node in Figure 3.3, in order to compute the generic output message, each check node performs the following computation:

$$m_j^c = 2\ \text{atanh}\prod_{\substack{i=1\\ i\neq j}}^{d_c}\tanh\frac{m_i^c}{2} \tag{3.10}$$

where m_j^c is the j-th output message and m_i^c is the i-th input message coming from the variable nodes. As in the variable node case, the message coming from the j-th edge is not used for the computation of the outgoing message in the j-th edge. The messages can be interpreted as LLRs of the bits associated with the variable nodes towards/from which the message is directed.

At the end of the decoding process, each variable node computes an output reliability value as follows:

$$m^v = m_0 + \sum_{i=1}^{d_v} m_i^v \tag{3.11}$$

where d_v is the degree of the node. In other words, the output reliability value of a codeword bit is the sum of all messages directed towards the corresponding variable node. As for the forward-backward (FB) algorithm, this step may be referred to as *completion*—see Chapter 2 for more details on the FB algorithm. From (3.6), an LLR referring to a binary random variable can be straightforwardly used to compute a MAP estimate of the random variable. In fact, if the sign of the LLRs is positive the probability of the random variable being equal to 0 is larger than the probability of the random variable being equal to 1. Vice versa, if the sign of the LLRs is negative the probability of the random variable being equal to 0 is smaller than the probability of the random variable being equal to 1. Thus, the signs of the LLRs in (3.11) are all is needed to obtain decisions on the codeword bits.

Summarizing, the Gallager C decoding algorithm comprises the following steps.

1. Compute all the reliability values for the symbols in the codeword. These values correspond, for each variable node, to the m_0 value in Figure 3.2.

2. Initialize to 0 all the messages coming from the check nodes.

3. Compute the variable nodes' output messages using (3.9).

4. Transfer the messages to the check nodes.

5. Compute the check nodes' output messages using (3.10).

6. Transfer the messages to the variable nodes.

7. Verify that a *stopping criterion*—described in the following paragraph—is met. If not, go to step 2.

8. Compute the final reliability values using (3.11).

A stopping criterion may be based on several possible events. The two most common are (i) the codeword bits that would be obtained after the completion step form a valid codeword, and (ii) a given maximum number of iterations is reached. It has been observed [41] that the Gallager C algorithm for LDPC codes has the interesting property of exhibiting particularly low probability of not detecting an erroneous decoding result, i.e., when the decoding process fails, the decoder is mostly aware of the failure. In fact, unlike turbo codes, the BP algorithm for LDPC codes operates on codeword bits rather than information bits. This is easily recognized by observing that the information bits (i.e., the information payload in the codeword) and the parity bits are dealt with in the same way. Neither convergence to the optimum codeword nor convergence to a codeword at all is guaranteed. It seems apparent that, whenever a decoding error occurs, the resulting decided bit sequence is not a codeword, in the sense that usually it does not satisfy (3.2).

In the above described algorithm, the structure of message passing is rigid: first, every variable node computes its messages; then, *all* the messages are passed to the check nodes, which in turn compute all the messages at their outputs. All the obtained messages are then sent back to the variable nodes. This scheduling is optimum when applied to a tree-shaped bipartite graph, i.e., a graph without cycles. If the graph has cycles, the algorithm becomes suboptimal and there may be some benefit in adopting other scheduling schemes. Another motivation for using other scheduling patterns, rather than the standard Gallager C algorithm, is to improve the computational and implementation efficiencies of circuits devoted to LDPC decoding.

In the following, we will refer to the set of variable node processors as *variable node detector* (VND) and to the set of check node processors as *check node detector* (CND). The LDPC decoding process can be seen as an iterative exchange of vector messages, referred to as *message sets*, between VND and CND.

3.4.2 Min-Sum Algorithm

The check node operation (3.10) in the log-domain relies on the computation of complex nonlinear functions, i.e., atanh(·) and tanh(·). The computation of (3.10) may be formulated in a recursive form as described in the following. Let $(m_1, \ldots, m_{d_c-1})$ denote the messages in the product of (3.10), where the superscript c has been omitted and consecutive indices have been adopted to simplify the notation. If we define the following recursion:

$$\begin{aligned} m_1^* &= m_1 \\ m_{i+1}^* &= 2\,\mathrm{atanh}\tanh\frac{m_i^*}{2}\tanh\frac{m_{i+1}}{2} \end{aligned} \tag{3.12}$$

then

$$m_j^c = m_{d_c-1}^* .$$

Note that, for any $x, y \in \mathbb{R}$

$$\begin{aligned} &2\mathrm{atanh}\left(\tanh\frac{x}{2}\tanh\frac{y}{2}\right) \\ &= \mathrm{sgn}(x)\mathrm{sgn}(y)\left(\min\{|x|,|y|\} + \log\frac{1+e^{-|x|-|y|}}{1+e^{-||x|-|y||}}\right) \end{aligned}$$

where the term

$$\log\frac{1+e^{-|x|-|y|}}{1+e^{-||x|-|y||}} \tag{3.13}$$

becomes small whenever $||x|-|y||$ is large. This result can be obtained by observing that, first, both $\tanh(\cdot)$ and $\mathrm{atanh}(\cdot)$ have odd symmetry and, therefore,

$$2\mathrm{atanh}\left(\tanh\frac{x}{2}\tanh\frac{y}{2}\right) = 2\mathrm{sgn}(x)\mathrm{sgn}(y)\mathrm{atanh}\left(\tanh\frac{|x|}{2}\tanh\frac{|y|}{2}\right) .$$

Expanding the right-hand side, one obtains:

$$\begin{aligned} 2\mathrm{atanh}\left(\tanh\tfrac{|x|}{2}\tanh\tfrac{|y|}{2}\right) &= \log\frac{1+\tanh(|x|/2)\tanh(|y|/2)}{1-\tanh(|x|/2)\tanh(|y|/2)} \\ &= \log\frac{1+e^{-|x|-|y|}}{e^{-|x|}+e^{-|y|}} \\ &= -\log\left(e^{-|x|}+e^{-|y|}\right) + \log\left(1+e^{-|x|-|y|}\right) \\ &= \min\{|x|,|y|\} - \log\left(1+e^{-||x|-|y||}\right) \\ &\qquad + \log\left(1+e^{-|x|-|y|}\right) \\ &= \min\{|x|,|y|\} - \log\frac{1+e^{-|x|-|y|}}{1+e^{-||x|-|y||}} \end{aligned}$$

where the well known identity

$$\log\left(e^a+e^b\right) = \max\{a,b\} + \log\left(1+e^{-|a-b|}\right)$$

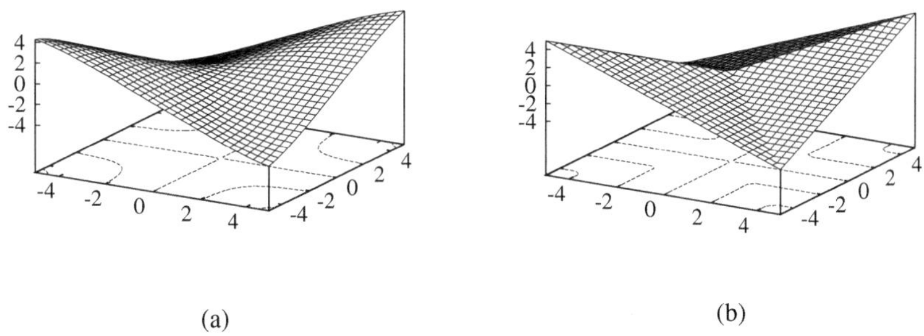

Figure 3.4: Comparison of the check node functions: (a) BP algorithm and (b) Min-Sum algorithm.

has been used. By neglecting the term (3.13) in the computation of the recursion (3.12), one obtains the following approximate check node operation:

$$m_j^c = \prod_{\substack{i=1 \\ i \neq j}}^{d_c} \operatorname{sgn}(m_i^c) \min_{i \neq j} \{m_i^c\}$$

which yields the so-called Min-Sum approximation of the BP algorithm.

In Figure 3.4.2, the check node operation in the LLR domain for a degree-3 check node is shown considering (a) the BP algorithm and (b) the Min-Sum algorithm. The x and y axes represent the two input LLR values and the z axis represents the output message in the LLR domain. One can observe the visual similarity of the two functions. In Figure 3.5, the difference between the exact function and its Min-Sum approximation is shown. One can observe that the highest difference is concentrated in the region where the two message values are close. If the input reliabilities to the variable nodes are characterized by high LLR values, as in the case of a good channel such as an AWGN channel with high SNR, the average difference between the Min-Sum approximation and the BP is expected to be small.

Min-Sum decoding has an interesting property that we will briefly discuss. Observe that, in the case of binary-input AWGN (BIAWGN) channel described in Example 1.1, the input LLR reliability value to the k-th variable node can

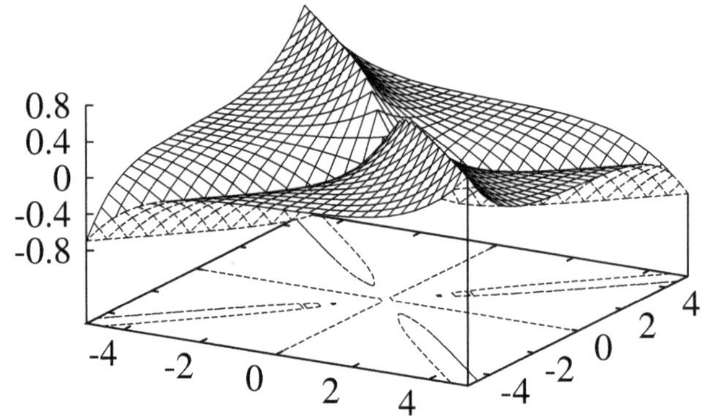

Figure 3.5: Difference between the BP and Min-Sum check node functions.

be computed as follows:

$$m_0 = \log \frac{\frac{1}{\sqrt{2\pi\sigma_w^2}} e^{-\frac{(y_k-1)^2}{2\sigma_w^2}}}{\frac{1}{\sqrt{2\pi\sigma_w^2}} e^{-\frac{(y_k+1)^2}{2\sigma_w^2}}} = \log e^{\frac{4y_k}{2\sigma_w^2}} = \frac{2y_k}{\sigma_w^2}$$

where y_k is the k-th received sample, i.e., signal plus noise, and σ_w^2 is the additive noise variance. In order to perform LDPC decoding with the above described BP algorithm, the only needed channel information consists of the knowledge of σ_w^2, which is used to compute the input LLRs. If in a variable node or check node Min-Sum operation all the messages are multiplied by a factor $\alpha > 0$, the resulting output LLR is multiplied by the same factor α. Therefore, if in the Min-Sum algorithm every input LLR is multiplied by α, the obtained messages, at every iteration and in every graph edge, change only by a constant factor α. In particular, the message sign does not change and, as a consequence, the bit decisions do not change.

By choosing $\alpha = \sigma_w^2/2$, one obtains a decoding algorithm that does not rely on the knowledge of the channel statistics, i.e., knowledge of σ_w^2 is unneeded and the input messages can be computed simply as

$$m_0 = y_k \, .$$

For this reason, the Min-Sum decoder is also known as *universal decoder* [42]. The performance of the Min-Sum decoder can be improved in a number of ways. In particular, the application of a correction factor and an offset in the check node output message may have beneficial effect on the decoder performance [43].

3.4.3 Alternative Decoding Algorithms

Several techniques have been proposed either to achieve better performance than that of BP decoding or to obtain low-complexity decoding. Linear programming (LP) decoding of LDPC codes has been proposed in [44, 45], where it is shown how to formulate the maximum likelihood (ML) decoding of LDPC codes as a LP problem. To make the use of an efficient LP algorithm feasible, the constraints on the solution must be (approximately) relaxed. Several improvements have been proposed to tighten the distance between approximate LP decoding and ML decoding. LP decoders have, in general, better performance than BP decoders. Nevertheless, their use as component blocks of the complex iterative receivers that will be investigated in the next chapters is difficult since they are conceived to perform *per-codeword* ML detection as opposed to *per-bit* MAP detection, which is better suited for iterative detection [46].

Other reduced complexity decoding techniques comprise bit-flipping techniques, which refer to graph-based algorithms with binary valued messages (see, e.g., [9] and [47] and references therein). Message quantization has been also investigated. The basis for performance analysis using quantized message passing algorithm for LDPC decoding may be found in [34].

3.5 Practical LDPC Code Design from Statistical Description

As usual, design techniques rely on performance evaluation techniques. The most effective LDPC code performance evaluation techniques will be discussed in Chapter 4 and are based on asymptotic analysis of LDPC code ensembles defined by their degree distributions. The use of these techniques in code design allows to optimize the degree distributions of the code, i.e., its statistical description. In order to use the code, it is necessary to construct a parity check matrix that satisfies, in addition to the obtained degree distributions, all the desired constraint, such as, for example, the codeword length.

A widely known LDPC construction algorithm is the progressive edge growth (PEG) algorithm [48, 49], which enables to design good LDPC codes characterized by high minimum cycle length, or *girth*, and based on a given degree distribution.

Most of the results obtained in this book will be based on LDPC codes constructed using the simple random graph construction algorithm described in the following.

The key goal is to obtain a code with, in order of priority:

1. a given codeword length N;
2. a girth larger than or equal to a given value γ;
3. a given rate R;
4. given degree distributions.

In general, it is not possible to satisfy the last two constraints. In fact, since R is generally a real number and the degree distributions have real coefficients, there may be cases where it is not possible to build a finite length code characterized by given code rate R and degree distributions. These two design constraints may need to be somehow relaxed, allowing for small roundings on the degree distributions coefficients.

In order to build a code we first compute, based on N and the variable node degree distribution, the number ℓ of edges in the graph

$$\ell = \frac{N}{\sum_i \frac{\lambda_i}{i}}$$

and the number v_i and c_i of degree-i variable and check nodes, respectively, for each i:

$$\begin{aligned} v_i &= \ell \frac{\lambda_i}{i} \\ c_i &= \ell \frac{\rho_i}{i} . \end{aligned}$$

The obtained values must be properly rounded in order to obtain an integer number of nodes for each degree. In addition, the rounding strategy for v_i must be chosen in order to obtain a total number of variable nodes equal to N, and the rounding of c_i must preserve the number of actual edges ℓ in the variable nodes.

To build the LDPC code graph, first the available variable and check nodes must be arranged in two separate groups. Each degree-i node of a specific kind,

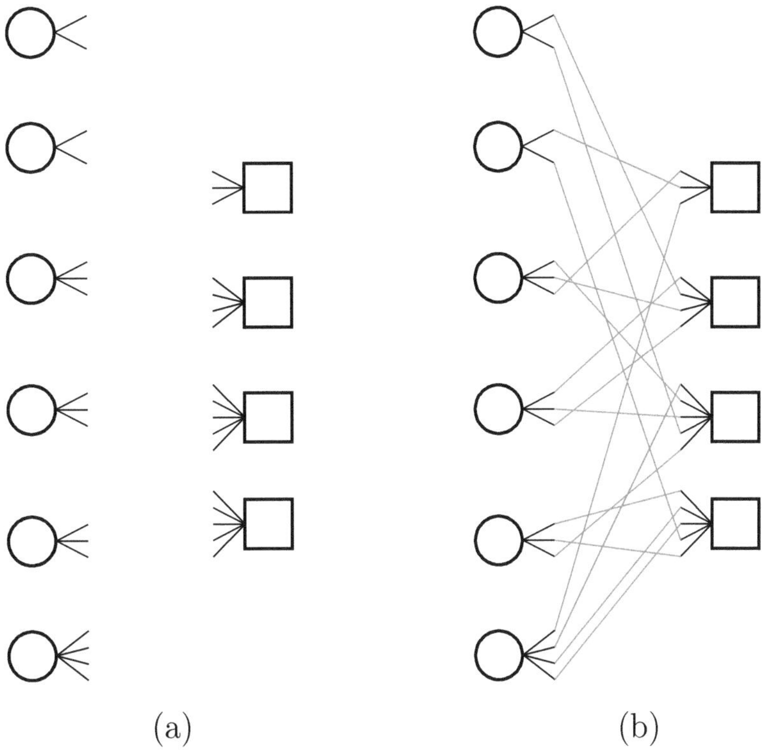

Figure 3.6: Simple code construction: (a) arranged nodes with empty sockets and (b) graph after all empty sockets have been connected.

either check or variable, has i "empty sockets," each ready to be connected to a socket of a node of the other kind. Figure 3.6 (a) shows the arranged nodes ready to be connected. After having prepared the nodes to be connected, the construction algorithm is as follows:

1. start from the first socket in the first variable node;
2. connect the current socket with a random socket in the check nodes;
3. verify that the new connection does not generate a cycle up to a given desired girth γ (by exploring the graph starting from the new edge up to a depth $\gamma - 1$);
4. if no new short cycle is generated pass to the next empty socket in the variable nodes and repeat starting from 2; else, delete the connection

and repeat from 2.

If the algorithm reaches a dead end, i.e., it is not possible to find an empty check node socket that does not generate a cycle of length longer than γ, all the connections are erased and the algorithm restarts from the beginning. If after a given number of trials the algorithm does not provide a complete graph, then the algorithm fails and returns no code at all. In this case, it may be useful to reduce the required girth or to increase the codeword length. Eventually, all empty sockets will be connected, as shown in Figure 3.6 (b).

3.6 Encoding Techniques for LDPC Codes

On the basis of the vector of information bits to be transmitted, encoding is the operation that selects, from the set of all codewords, i.e., the codebook, one corresponding codeword. In general, LDPC code encoding can entail significant complexity due to the random code structure. In the following subsections, some encoding techniques are described that may be applied to any generic LDPC code.

3.6.1 Encoding by Matrix Multiplication

In several contexts involving binary coding, it is useful that the information bits explicitly appear in the associated codeword. This allows, for example, to avoid costly decoding when the communication system is operating on a particularly favorable channel, introducing negligible uncertainty on the received data sequence. Codes satisfying this condition are known as *systematic*. The part of codeword replicating the data bits is known as the *systematic portion* of the codeword. In convolutional encoding, the systematic portion is often interleaved with the non-systematic part (also known as *parity bits*). In this book, we will address systematic LDPC codes whose systematic part consists of the first bits of the codeword.

Since LDPC codes are linear block codes, they can be encoded using the corresponding $N \times K$ generation matrix G. In particular, the codeword $\boldsymbol{x}$ can be obtained as follows:

$$\boldsymbol{x} = G\boldsymbol{a}$$

where $\boldsymbol{a} = (a_1, \ldots, a_K)^T$ is the vector of K information bits to be encoded. Since the code is systematic, the generation matrix must be of the form

$$G = \begin{pmatrix} I_K \\ \cdots \\ P \end{pmatrix}.$$

where P is a $(N-K)\times K$ matrix that generates the parity bits of the codeword. The encoded vector $\boldsymbol{x} = G\boldsymbol{a}$ is in the form

$$(a_1, \ldots, a_K, p_1, \ldots, p_{N-K})^T$$

where the vector of parity bits is

$$\boldsymbol{p} = (p_1, \ldots, p_{N-K})^T = P\boldsymbol{a}\,.$$

The matrix $\tilde{H}$ defined as

$$\tilde{H} = \left(P \vdots I_{N-K}\right)$$

is a parity check matrix for the systematic code generated by G, meaning that a vector $\boldsymbol{x}$ is a codeword if and only if

$$\tilde{H}\boldsymbol{x} = 0\,.$$

This fact can be easily recognized since any bit vector $\boldsymbol{x} = (x_1, \ldots, x_N)$ is a codeword only if the last $N-K$ bits $\hat{\boldsymbol{p}} = (x_{K+1}, \ldots, x_N)$ are equal to the parity bit vector $\boldsymbol{p}$ generated by the first K codeword bits as follows:

$$\boldsymbol{p} = P(x_1, \ldots, x_K)^T\,.$$

It is easily recognized that

$$\tilde{H}\boldsymbol{x} = \left(P \vdots I_{N-K}\right)(x_1, \ldots, x_K, x_{K+1}, \ldots, x_N) = \boldsymbol{p} + \hat{\boldsymbol{p}}$$

which is equal to 0, using boolean algebra, if and only if

$$\boldsymbol{p} = \hat{\boldsymbol{p}}$$

i.e., if and only if $\boldsymbol{x}$ is a codeword.

The generation matrix G is completely defined by its sub-matrix P, which, in turn, can be computed by transforming a given parity check matrix H into its $\tilde{H}$ form, by means of simple row operations (i.e., substitution of a row with the linear combination of itself and other rows).

Given a parity check matrix H, in order to guarantee the existence of a corresponding systematic code, the rightmost $(N-K)\times(N-K)$ sub-matrix of H must be non singular. If this condition does not hold, however, observe that, if H has maximum rank, there exists a subset of $N-K$ columns of H that forms a non-singular matrix. Therefore, it is possible to obtain a parity check matrix corresponding to a systematic code by applying a permutation of the columns of H that places those columns in the rightmost positions.

Although H is sparse by definition, the sub-matrix P of G is, in general, dense. This means that, discarding the identity matrix part, in order to store G, at least

$$K(N-K) = K^2\left(\frac{1}{R}-1\right)$$

bits of memory are needed. Since a typical LDPC codeword size ranges from 10^3 to 10^4 it means that the memory for storing G for, e.g., a rate 1/2 code (i.e., $N-K=K$), is at least on the order of 10^6 bits.

The matrix multiplication operation, accounting only for the dense part of G, requires $K(N-K) = K^2(1/R-1)$ multiplications (boolean AND operations) and $(K-1)(N-K) = (K^2-K)(1/R-1)$ additions (boolean EXOR operations). The encoding complexity is therefore *quadratic* in the number of information bits per codeword or, given the code rate, in the length of the codeword. It is also an increasing function of the code rate. The $P\boldsymbol{a}$ matrix multiplication can be intrinsically done in parallel, by performing several scalar products. This may be useful if speed or latency are critical issues, although it requires a specific implementation.

3.6.2 Recursive Encoding and Structured Codes

Since BP decoding has linear computational cost in the block length, for long enough codes, encoding may become more computationally expensive than decoding itself. To overcome this problem, several solutions have been proposed. A possible technique is to design highly structured codes. This approach generated a thriving field of research that lead to the invention of several coding techniques. A particularly interesting possibility is to build LDPC codes that are quasi-cyclic, thus enabling low-complexity encoding [50–52]. Another example is given by [53], where turbo-Gallager codes are introduced. They consist of a class of turbo codes which can be effectively decoded by means of a standard LDPC code decoder and, simultaneously, are characterized by the linear complexity encoding property of turbo codes. A family of codes which enable efficient recursive encoding is proposed in [54], with particular emphasis on digital subscriber line applications. Due to their highly structured parity check matrix, the proposed codes enable the use of a very simple circuit for encoding purposes. Besides possible simplification of the encoding procedure, structured codes might have other interesting benefits including

- the possibility of an algorithmic description of the parity check matrix, thus avoiding the storage of the whole matrix;

- a more efficient interconnection of the processing blocks involved in the LDPC code decoding: an unstructured LDPC code requires the capability of a generic interconnection structure which might lead to unsolvable difficulties in the design of the system hardware;

Another possible technique to obtain efficient encoding is proposed in [55], where it is shown how to exploit the sparseness of the parity check matrix to obtain quasi-linear encoding complexity. The proposed technique exhibits a one-to-one correspondence with the decoding process for an erasure channel using the same code. The procedure is recursive, i.e., all the parity bits but a small fraction are obtained using a recursion, whereas the remaining bits are obtained by means of a matrix multiplication.

It is interesting to note that, if the parity check matrix can be put in lower triangular form in its rightmost part by means of a row permutation, then a BP decoding process, initialized with zero LLRs for the parity bits part and the value $1 - 2a_k$ in the k-th systematic bit, completely recovers all the parity bits after a sufficiently large number of iteration (at most $N-K$). This means that, if the parity check matrix fulfills the previously given condition, the decoder can be used for encoding as well. If the parity check matrix does not fulfill the requirements for exact *encoding by decoding*, then at the end of the encoding process there will be a fraction of parity bits that remain uncertain. In principle, it could be possible to use the partial encoding result to compute a signal to be transmitted. Depending on the strategy used this would result in a performance loss, due to a *suboptimal encoding*.

3.7 Concluding Remarks

In this chapter, an overview of LDPC codes has been given. We remark that, on the basis of the structure and properties of LDPC codes, a huge number of new code families have been proposed in several works. In this book, we focus on generic irregular LDPC codes since we will use an LDPC code as a component block of a more complex system tailored for transmission on specific channels. In the next chapter, an overview of the most important performance analysis techniques for iterative detection schemes will be given, with particular emphasis on LDPC coded schemes.

Chapter 4

Performance Analysis Techniques

4.1 Introduction

Efficient system design requires the use of tools for evaluating the performance of a given system. Based on a performance analysis method, design algorithms can be constructed, for example, with a specific performance optimization goal. The ideal analysis tool should be (i) fast, (ii) accurate, and, possibly, (iii) highly informative, i.e., it should provide a comprehensive description of the system under analysis. However, exact analytical evaluation of the relevant statistical parameters characterizing a realistic communication system is, usually, unfeasible. Therefore, it is necessary to resort to approximate evaluation tools. A universal technique satisfying the requirements (ii) and (iii) is given by a Monte Carlo simulation-based analysis of the system performance. In particular, this simulation method allows to collect all needed system statistical parameters with the desired accuracy. Unfortunately, system simulation is usually a computationally intensive task. This makes this technique not appealing for automated system parameter optimization (or design), which usually calls for repeated analyses. Nevertheless, Monte Carlo simulation is an invaluable tool suited for accurately testing the performance of a complex system. Iterative receiver schemes can also be investigated through a number of approximate tools which exploit their internal structures. For example, *density evolution* and extrinsic information transfer (EXIT) charts admit efficient numerical implementation, since they are characterized by a limited computational complexity.

In this chapter, the main numerical performance analysis tools useful for

M. Franceschini, G. Ferrari and R. Raheli, *LDPC Coded Modulations*,
DOI: 10.1007/978-3-540-69457-1_4, © Springer-Verlag Berlin Heidelberg 2009

the analysis and design of low-density parity-check (LDPC) codes for coded modulations will be described. In Section 4.2, Monte Carlo techniques are introduced and some considerations on their correct implementation and use are drawn. In Section 4.3, the density evolution method for analysis of iterative decoders is described. In Section 4.4, EXIT charts are introduced. Section 4.6 concludes this chapter.

4.2 Monte Carlo System Simulation

By system simulation, we refer to the practice of reproducing a set of signals which are statistically equivalent to those found in the actual system under analysis. To this end, the system is decomposed into its component blocks. We assume that the system may be decomposed into a network of blocks characterized by deterministic behavior and driven by inputs which may be either deterministic or stochastic. This assumption is quite general and may be applied in almost all scenarios of interest.

The deterministic blocks of the system are reproduced by implementation of the corresponding (embedded) signal processing algorithms or by numerical solution of the input-output equations of their mathematical model. The stochastic input signals may be generated in two different ways.

The first is to implement a physical device comprising a controlled and precisely known noise source, such as, e.g., amplified thermal noise, which is used to obtain a signal whose statistics are similar to those of the stochastic input to the system. In this case, the fact that the noise source parameters are not perfectly known is a possible cause of mismatch between the simulated system and the actual one.

The second, and most relevant for our purposes, technique for generating the input stochastic signals is by means of a pseudo-random number generator (PRNG). PRNGs are recursive algorithms which may be viewed as autonomous systems whose initial state is referred to as *seed* and whose output, interpreted as the realization of an ergodic and stationary process, implies statistical properties of the process such as, e.g., independent and identically distributed (i.i.d.) outputs, uniform distribution of the output sample over the integers within a given range, etc. Obviously, since the output is actually a deterministic sequence, the statistical description of the sequence is only an approximation of the desired one. Nevertheless, the output sequence is plausible for the stochastic model that the PRNG tries to emulate. The "randomness" of the sequence can be evaluated using a number of statistical evaluation mathematical tools. In order to obtain stochastic processes with

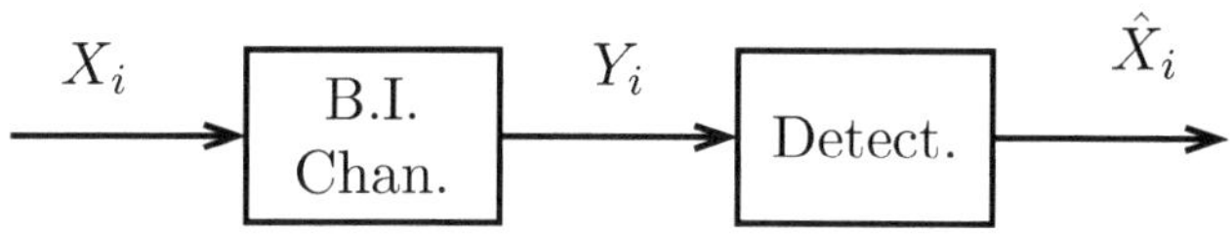

Figure 4.1: Schematic diagram of a uncoded binary-input (BI) simulator.

the desired statistical description, the output of the PRNG can be properly processed applying standard methods [56].

It is important to note that not every PRNG may be suited for the specific scope of a problem. On the other hand, in general, it is not necessary to use a true—as opposed to pseudo—random number generator. Every simulation scenario has its own characteristics. For example, almost every PRNG has a periodic behavior, i.e., the sequence repeats itself after a finite number of samples. Some applications may require a very long period in order to guarantee accurate estimation of the relevant statistical parameters. Other applocations, such as, for example, the generation of information bits in an uncoded binary-input (BI) memoryless channel simulator, can achieve the same or better results than those obtained with a true random number generator, with a PRNG with period as short as 2. This is considered in the following example.

Example 4.1 *Simple PRNGs might be better than true RNGs*

We now compare the quality of BER estimate using a real random input data sequence and one generated using a PRNG with cycle of length 2. Showing that, even if the real communication system will operate on a random data sequence, the use of a PRNG leads to better estimate of the BER of that system. Consider a simulator for the uncoded transmission scheme in Figure 4.1, where the input sequence $\{X_i\}$ is a random sequence of i.i.d. bits $\{X_i\}$ with $P\{X_i = 1\} = P\{X_i = 0\} = 1/2$. The symbol Y_i output by the BI channel at epoch i depends only on X_i, i.e., the channel is memoryless. The sequence is detected by means of a detector device which outputs estimates of the transmitted bits on a bit-by-bit basis. For example, choose the maximum *a posteriori* probability (MAP) symbol decision rule, i.e., choose the bit $\hat{X}_i$ whose conditional a posteriori probability, given the received Y_i, is highest. We assume that the conditional probability of error $P_{e,0}$ associated with the transmission of a 0 might be different from the conditional probability of error $P_{e,1}$ associated with the transmission of a 1.

We use Monte Carlo simulation for measuning the bit error rate (BER) of the system. We generate N bits to be transmitted $X_1, \ldots, X_N$ and send them through the channel obtaining N output samples $Y_1, \ldots, Y_N$. As BER estimator we consider the sample average of the error indicator function:

$$\hat{P}_e = \frac{1}{N} \sum_{i=1}^{N} 1\{\hat{X}_i \neq X_i\}$$

where the *indicator function* $1(\mathcal{A})$ of an event $\mathcal{A}$ is given by the following definition:

$$1(\mathcal{A}) = \begin{cases} 1 & \text{if } \mathcal{A} \text{ is true} \\ 0 & \text{else.} \end{cases}$$

In other words, $\sum_{i=1}^{N} 1\{\hat{X}_i \neq X_i\}$ is the total number of errors in N trials.

We now characterize the variance of the BER estimator considering the two cases of a real random number generator and a PRNG.

First we measure probability of error using real random number generators. The bits to be transmitted are therefore N i.i.d. equiprobable random bits $X_1, \ldots, X_N$. The mean of the estimator is

$$\mathrm{E}\{\hat{P}_e\} = \frac{1}{N} \sum_{i=1}^{N} \mathrm{E}\{1\{\hat{X}_i \neq X_i\}\} = \frac{1}{N} \sum_{i=1}^{N} P_e = P_e$$

where the fact that $\mathrm{E}\{1(\mathcal{A})\} = \Pr\{\mathcal{A}\}$ has been used and $\Pr\{\mathcal{A}\}$ denotes the probability of event $\mathcal{A}$. As a consequence, the estimator is *unbiased* and its variance is

$$\begin{aligned}
\mathrm{Var}\{\hat{P}_e\} &= \frac{1}{N^2} \sum_{i=1}^{N} \mathrm{Var}\{1\{\hat{X}_i \neq X_i\}\} \\
&= \frac{1}{N^2} \sum_{i=1}^{N} \mathrm{E}\{1^2\{\hat{X}_i \neq X_i\}\} - \mathrm{E}^2\{1\{\hat{X}_i \neq X_i\}\} \\
&= \frac{1}{N^2} \sum_{i=1}^{N} P_e - P_e^2 \\
&= \frac{P_e - P_e^2}{N}
\end{aligned}$$

where $\mathrm{Var}\{\cdot\}$ denotes the variance of a random variable.

Assume now that we can accurately simulate the exact statistical channel distribution (e.g., by *using* the actual channel) but we send alternatively 1 and 0, i.e., $X_{2i-1} = 1$ and $X_{2i} = 0$, $i \geq 1$. This corresponds to the use of a particular PRNG with periodicity equal to 2 to generate the $\{X_i\}$ sequence.

As before, we generate N (even) samples and compute the mean and variance of our estimator applied to the newly obtained sequence. Since it holds that

$$
\begin{aligned}
\mathrm{E}\{\hat{P}_e\} &= \mathrm{E}\left\{\frac{1}{N}\sum_{i=1}^{N} 1\{\hat{X}_i \neq X_i\}\right\} \\
&= \mathrm{E}\left\{\frac{1}{N}\left[\sum_{i=0}^{N/2-1} 1\{\hat{X}_{2i+1} \neq X_{2i+1}\} + \sum_{i=1}^{N/2} 1\{\hat{X}_{2i+2} \neq X_{2i+2}\}\right]\right\} \\
&= \frac{1}{N}\left[\sum_{i=0}^{N/2-1} \mathrm{E}\{1\{\hat{X}_{2i+1} \neq X_{2i+1}\}|X_{2i+1} = 1\}\right. \\
&\quad \left. + \sum_{i=1}^{N/2} \mathrm{E}\{1\{\hat{X}_{2i+2} \neq X_{2i+2}\}|X_{2i+2} = 0\}\right] \\
&= \frac{1}{N}\left[\frac{N}{2} P\{\hat{X}_{2i+1} \neq X_{2i+1}\}|X_{2i+1} = 1\}\right. \\
&\quad \left. + \frac{N}{2} P\{\hat{X}_{2i+2} \neq X_{2i+2}\}|X_{2i+2} = 0\}\right] \\
&= \frac{1}{2} P_{e,1} + \frac{1}{2} P_{e,0} \\
&= P_e
\end{aligned}
$$

it follows that the estimator is *unbiased*. Moreover, its variance is

$$
\begin{aligned}
\mathrm{Var}\{\hat{P}_e\} &= \frac{1}{N^2}\sum_{i=1}^{N} \mathrm{Var}\{1\{\hat{X}_i \neq X_i\}\} \\
&= \frac{1}{N^2}\left[\sum_{i=0}^{N/2-1} \mathrm{Var}\{I\{\hat{X}_{2i+1} \neq X_{2i+1}\}\right. \\
&\quad \left. + \sum_{i=0}^{N/2-1} \mathrm{Var}\{1\{\hat{X}_{2i+2} \neq X_{2i+2}\}\}\right] \\
&= \frac{1}{N^2}\left[\frac{N}{2}(P_{e,1} - P_{e,1}^2) + \frac{N}{2}(P_{e,0} - P_{e,0}^2)\right] \\
&= \frac{1}{N}\left(\frac{P_{e,1} + P_{e,0}}{2} - \frac{P_{e,1}^2 + P_{e,0}^2}{2}\right) \\
&\leq \frac{P_e - P_e^2}{N}
\end{aligned}
$$

where $P_{e,1} = P\{\hat{X}_{2i+1} \neq X_{2i+1} | X_{2i+1} = 1\}$ and $P_{e,0} = P\{\hat{X}_{2i+2} \neq X_{2i+2} | X_{2i+2} = 0\}$ and the last passage is due to the fact that

$$P_e^2 = \left(\frac{P_{e,1} + P_{e,0}}{2}\right)^2 \leq \frac{P_{e,1}^2 + P_{e,0}^2}{2}.$$

This means that the resulting variance of the estimator is better than or equal to that obtained with random bits, i.e., using realistic data. It turns out that if the channel is known to be symmetric, i.e., if $P_{e,0} = P_{e,1}$, as they do in the case of the BI additive white Gaussian noise (BIAWGN) channel investigated in Example 1.1, the two variances coincide.

In performing a system analysis through Monte Carlo simulation, one should carefully consider which parameters are to be estimated. This is important, since a simulation, in general, produces a large amount of data. It is usually not possible nor efficient to completely store the produced data and, afterwards, proceed with a batch processing. Typically, depending on the particular statistical parameter to be evaluated, only a small fraction of the produced data has to be *stored*, although the whole produced data may sometimes need to be *processed*. As a simple but meaningful example, consider the transmission system analyzed in the above example. In order to estimate the BER, one could store the sequences $\{Y_i\}$, $\{X_i\}$, and $\{\hat{X}_i\}$, then perform batch processing on the sequences and compute both the estimate $\hat{P}_e$ and and an estimate of its variance. This would require a memory of size $O(N)$.[1] Obviously, the same can be done with a memory as small as $O(1)$. In fact, the processing could be done by updating an *error counter* on a sample-by-sample (or bit-by-bit) basis, i.e., if $\hat{X}_i \neq X_i$ then increase the *error counter* by one. The estimate is obtained by dividing the number of errors by N. Its variance can be estimated, in a similar way, using a single update variable.

One should always remember that every estimate based on the output of a simulation is, as a matter of fact, a *measure* which is subject to error. Assume to simulate the transmission of N bits by a communication system. By comparing the decided bits at the output of the receiver with the transmitted ones, one discovers that there are X errors. Without any other knowledge, the most reasonable estimate of the BER $\tilde{P}_\mathrm{e}$ characterizing the system is

$$\tilde{P}_\mathrm{e} = \frac{X}{N}.$$

[1]We say that a quantity $g(N)$ is "on the order of" $f(N)$ or $O(f(N))$, if the limit

$$\lim_{N\to\infty} \frac{f(N)}{g(N)}$$

exists and is greater than 0.

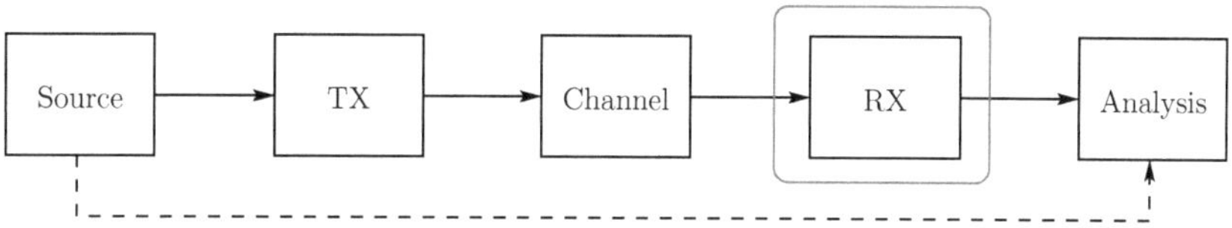

Figure 4.2: Block diagram of a Monte Carlo analysis system.

This estimator corresponds to the sample mean of the error indicator, and has good properties in several scenarios of practical interest. For example, if the communication system processes are stationary, it turns out that

$$\mathrm{E}\{\tilde{P}_{\mathrm{e}}\} = P_{\mathrm{e}}$$

where P_{e} is the true system BER. Note that every estimator is characterized by its own distribution. This allows to compute the uncertainties that should characterize every well done measure. For more details on this important topic we refer the reader to [56] and [57].

An important property of system simulation as analysis tool is that, if properly implemented, it is robust against possible implementation/programming errors. In Figure 4.2, a block diagram of a Monte Carlo simulation-based analysis system is shown. Each block, source, transmitter (TX), channel, receiver (RX), and statistical analysis, denote a "separate" software unit that may be implemented exploiting the most convenient paradigms of the adopted language or toolkit. By "separate" software unit we mean that it does not rely on nor may access other data besides those passed through the arrow connection, which carry only the sampled signal through the system. The receiver block is highlighted to emphasize that it is the block where the biggest design effort is spent. The other blocks, except for the transmitter which may make use of sophisticated signal processing algorithms, are usually very simple. If every other processing block but the receiver is known to be working correctly, a Monte Carlo simulation guarantees the following interesting robustness property: *if the receiver implementation is defective, i.e., it comprises one or more implementation errors, there cannot be false improvements of the system performance.* In other words, although defective, the implemented receiver algorithm can be considered as a real algorithm working on real data and the computed performance corresponds to the performance of a receiver using the buggy algorithm.

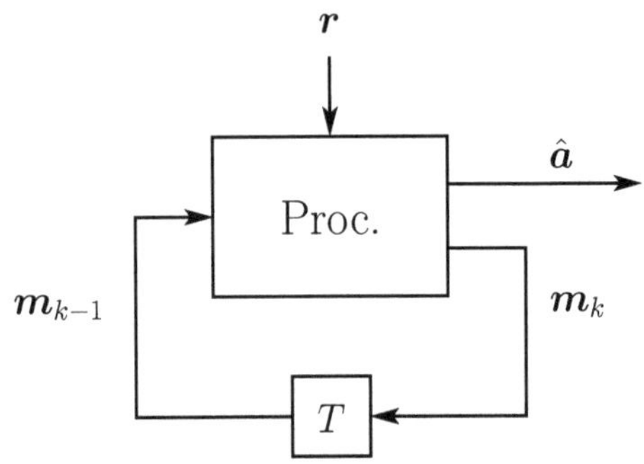

Figure 4.3: Illustrative diagram of a generic iterative detection scheme.

4.3 Density Evolution

Consider a generic iterative detection algorithm. The transmitted data sequence $\boldsymbol{a}$ produces a received vector $\boldsymbol{r}$ which is the input to the iterative detection scheme. A possible representation of a generic iterative algorithm is shown in the diagram in Figure 4.3. The receiver observes the vector $\boldsymbol{r}$ and generates a first vector of messages $\boldsymbol{m}_0$. The iterative process begins and at each iteration the previously obtained message set $\boldsymbol{m}_{k-1}$ is processed by a deterministic function which, on the basis of $\boldsymbol{m}_{k-1}$ and $\boldsymbol{r}$, computes the next message set $\boldsymbol{m}_k$. Note that, although the input vector $\boldsymbol{r}$ is random, the processing block operations are deterministic. At the end of the iterative process, e.g., after ℓ iterations, the output $\hat{\boldsymbol{a}}$ is computed as a function of the last message set $\boldsymbol{m}_\ell$ and of the input $\boldsymbol{r}$.

A complete statistical characterization of the iterative process would require to compute how the conditional joint probability density function (pdf) $p(\boldsymbol{m}_k|\boldsymbol{a})$ evolves as a function of k, and considering all possible transmitted data sequences $\boldsymbol{a}$. This is, however, impractical since the number of possible data sequences is an exponential function of the data sequence length and the size of the vector $\boldsymbol{m}_k$ is typically large. An LDPC code decoder would comprise a number of messages equal to the number of edges in the LDPC code's bipartite graph, which in practical applications, as a rule of thumb, may span from 2 to 6 times the codeword length. This implies that such an analysis of a practical LDPC code would require to compute the evolution of a joint pdf of about 10^4-10^5 random variables (RVs).

A technique which can be used to effectively analyze iterative decoding schemes is the so-called *density evolution* [34, 58]. This technique is based on a single-letter analysis of the evolution of the messages in an iterative decoding algorithm. In other words, the analysis focuses on the computation of the output distribution of a single message $m_k^{(i)}$. The message $m_k^{(i)}$ is a deterministic function of a subset $m_{k-1}^{(j_1)}, \ldots, m_{k-1}^{(j_{d_i})}$ of d_i elements of the vector of messages $\boldsymbol{m}_{k-1}$ at the output of the processing block at the iteration $k-1$:

$$m_k^{(i)} = f_i(m_{k-1}^{(j_1)}, \ldots, m_{k-1}^{(j_{d_i})}) . \tag{4.1}$$

The pdf of $m_k^{(i)}$ is a function of the joint distribution of the input messages involved in the computation:

$$p_{m_k^{(i)}} = \Psi_i(p_{m_{k-1}^{(j_1)}, \ldots, m_{k-1}^{(j_{d_i})}}) . \tag{4.2}$$

In general, the function $\Psi_i(\cdot)$ depends on the index i of the considered output message and can rarely be computed in closed form. Nevertheless, numerical approaches can be followed, for example by sampling the pdf and representing the messages with discrete RVs.

The discretization of the messages can be limited to the analysis purpose or, as practical scenarios call for low complexity implementations, could be the intrinsic way the receiver operates. In other words, the receiver may operate on discrete messages, e.g., the representation of the messages could be limited to 4 bits, in which case the messages belong to a 16 elements set. In a quantized scenario with M-level messages, the pdfs become probability mass functions (pmfs) which can be represented by M-element real vectors. The resulting pmf evolution function Ψ is

$$\Psi : \mathbb{R}^{M^{d_i}} \mapsto \mathbb{R}^M .$$

Observing (4.1) and (4.2), one can immediately notice two issues:

i. if at the k-th iteration the joint pdf of the messages $m_{k-1}^{(j_1)}, \ldots, m_{k-1}^{(j_{d_i})}$ is needed, the computation of the marginal pdf $p_{m_k^{(i)}}$ only does not allow to proceed iterating the algorithm;

ii. even if the pdf $p_{m_k^{(i)}}$ is available for every k, how can one understand if the decoding algorithm is converging, i.e., if the BER is approaching 0?

A solution for the first issue is to assume conditional independence of the messages $m_{k-1}^{(j_1)}, \ldots, m_{k-1}^{(j_{d_i})}$. In this case, (4.2) becomes

$$p_{m_k^{(i)}} = \Psi_i(p_{m_{k-1}^{(j_1)}}, \ldots, p_{m_{k-1}^{(j_{d_i})}}) \tag{4.3}$$

which entails significant simplification of the problem, as can be seen in the discrete message case with M-level messages. In this case, the pmf evolution function Ψ is

$$\Psi : \mathbb{R}^{Md_i} \mapsto \mathbb{R}^M .$$

Although, in general, this would be an approximation, this assumption holds exactly in the case of LDPC codes if the iteration number k is smaller than the girth of the graph.

Consider now the second issue, i.e., how to track the actual convergence of the decoding system based on the evolution of the pdfs $p_{m_k^{(i)}}$. At the final iteration, the processing block outputs a vector $\hat{\boldsymbol{a}}$, which is an estimate of the transmitted data sequence, computed as a function of the last message set $\boldsymbol{m}_\ell$ and of the received observable sequence $\boldsymbol{r}$. Therefore, the output value of a symbol $\hat{a}_l$ is a function $g_l(\boldsymbol{m}, \boldsymbol{r})$. A corresponding pmf $p_{\hat{a}_l}$ can be computed as a function of the pdf of $\boldsymbol{m}$. Once obtained the pmf of $\hat{a}_l$ and given that the analysis is done assuming a particular transmitted sequence, it is easy to compute the corresponding probability of error.

Another common assumption done in a density evolution analysis of LDPC codes is that of considering a common distribution equal for all the input messages. In other words, this corresponds to assuming that the input pdfs are all equal:

$$p_{m_{k-1}^{(j_1)}} = \cdots = p_{m_{k-1}^{(j_{d_i})}} \triangleq p_{m_{k-1}} .$$

Since, at each step, $\{p_{m_k^{(i)}}\}$ are computed and generally vary for different values of i, this requires *an additional step* after the computation of each relevant output pdf $p_{m_k^{(i)}}$. In this additional step, usually, averaging of all output pdfs is performed for computing the pdf p_{m_k} used as input pdf at the (next) iteration $k+1$:

$$p_{m_k} = \frac{1}{L} \sum_{i=1}^{L} p_{m_k^{(i)}}$$

where L is the number of messages, i.e., the length of $\boldsymbol{m}_k$.

The assumption of a unique distribution for all input messages holds exactly in the case of *regular* LDPC codes [58], assuming that the all-zero sequence has been transmitted (present in all LDPC codebooks). In general, in a communication system, the performance depends on the particular transmitted codeword. However, the sum-product (SP) algorithm has a *symmetry property* that allows to state that the statistical description of the messages does not depend on the particular codeword. Therefore, the use of density evolution leads, in this case, to exact results. For more details on density evolution techniques, which in general can be applied to a wide variety of message passing decoders, we refer the interested reader to [58].

We point out that the main disadvantage of this technique is that analytical derivation of the message pdf evolution function $\Psi(\cdot)$ is seldom feasible (with some important exceptions—see, for example, [13]). On the other hand, a statistical evaluation of the pdf evolution would need intensive Monte Carlo simulations, thus limiting the computational efficiency of this analysis technique.

4.4 EXIT Charts

Whenever its underlying assumptions hold, density evolution leads to a complete statistical characterization of the decoder, although it requires the capability of efficiently computing the evolution of an entire pdf (or probability mass function, pmf, if the message set is finite). Another possible solution is to track the evolution of some statistical function of the message set. This could lead to great simplifications. For example, in [59], the authors use a real valued function of the pdf of message sets, which is defined as an equivalent signal-to-noise ratio (SNR), as a means for characterizing the input-output relation of a processing block (the forward-backward, FB, algorithm for a particular convolutional code, in this case). The input SNR/output SNR relation of each component block of the iterative receiver is then used for predicting the convergence behavior of the receiver.

A statistical parameter, function of the message distribution, which allows to obtain good accuracy through this analysis method is the average mutual information (MI) between the generic transmitted codeword bit and the generic message in the decoder referring to that particular bit. In a SP decoder or, more generally, in a message passing decoder for LDPC codes, a message is said to *refer to a codeword bit* if it is originated from or is directed towards the variable node corresponding to that particular bit.

There is a number of possible techniques which can be used to compute the average MI of the messages. If every message m in the set has the same known distribution $p(m|a)$, where a is the corresponding bit, then the MI can be computed as usual:

$$I = \sum_a \int p(m|a)P(a) \log \frac{p(m|a)}{p(m)}\, dm. \tag{4.4}$$

However, reality is usually more complicated: (i) the distributions of the messages are not equal and (ii) the message pdf is not known. In the first case, a possible definition of the MI is the following:

$$I = \frac{1}{L} \sum_{i=1}^{L} I_i \tag{4.5}$$

where I_i is the MI between the i-th message and the corresponding bit, and L is the total number of messages.

In the second case, i.e., whenever the distribution is difficult to compute, several approximate techniques may be used. In particular, in some circumstances, it may be convenient to use a Monte Carlo approach to obtain an estimate of the MI. This can be done by setting up a simulator and generating sets of sample messages which are quantized and used to obtain a histogram approximation of the pdf $p(m|a)$. The MI can therefore be computed based on the histogram approximation and using (4.4).

The MI has a practical meaning: it quantifies, in terms of bits, how much information about a particular codeword bit a given message carries. In practice, whenever the MI is equal to 1, the bit is reliably recovered. This can be easily shown: in fact, if A is a uniformly distributed binary random variable, $H(A) = 1$. Assuming $I(A;M) = 1$, a functional relation exists between M and A. In fact:

$$\begin{aligned} I(A;M) &= H(A) - H(A|M) \\ &\Downarrow \\ 1 &= 1 - H(A|M) \\ &\Downarrow \\ H(A|M) &= 0 \end{aligned}$$

which implies that, given M, A is known, i.e., it is a function of M.

4.4.1 EXIT Curves and EXIT Charts

At this point, one could observe that although it is possible to track the evolution of the MI of the message sets during the iteration process, the MI is still a function of the underlying pdf of the messages and, therefore, little advantage may be obtained in this way. However, empirical observation shows that if a message set characterized by an MI I' is used as input to a given processing block, the MI I'' of the output message set depends almost only on I' and has little dependence on the particular pdf of the input messages. Each processing block can therefore be characterized by the relation between the input message set MI and the output message set MI. Since, usually, only extrinsic information is exchanged between processing blocks, this the plot of this relation is referred to as *extrinsinc information transfer (EXIT) curve*. EXIT curves are used to study convergence of recursive (iterative) detection/decoding algorithms by means of graphs usually referred to as *EXIT charts*. EXIT chart-based analyses allow to predict the system performance with a significantly lower computational burden with respect to the use of density evolution or standard computer simulations employed to evaluate the BER performance of iterative decoders [60, 61].

The assumption of independence on the particular message set pdf is an approximation, although it turns out to be an effective one. A more rigorous statement is that, given an input MI, the output MI is bounded within a range of possible values whose extremes are functions of the input MI. This range is, in practical applications, reasonably small. A thorough treatment of this topic, usually referred to as *information combining*, can be found in [62–64].

A remark is worthwhile at this point. We said that the processing block modifies the MI. In particular, we wish the MI to increase, at each iteration, in order to approach the value 1 as closely as possible. One can observe, however, that by the data processing inequality [1], it is not possible to increase the MI by means of data processing. In fact, the MI we refer to is not the MI between the codeword bits and the message set, i.e., the MI between two vectors. The MI used in EXIT charts is the MI between a message and its corresponding codeword bit, averaged over all messages in the message set. This means that the statistical dependence between different messages is purposely neglected. The processing block can therefore exploit this dependence in order to increase the EXIT chart MI.

4.4.2 SISO Detectors and EXIT Charts

A processing block computing bit reliabilities based on (i) a set of constraints, (ii) an optional set of observations from the channel and (iii) some input *a priori* information, is usually referred to as soft-input soft-output (SISO) detector or SISO module [65–68]. In the following chapters, we will use SISO modules in systems employing differentially-encoded phase shift keying (DE-PSK) and DE quadrature amplitude modulation (DE-QAM) transmission over an AWGN channel, DE-PSK with noncoherent detection, and PSK transmission through a channel affected by ISI.

Since we will focus on binary coding techniques, we assume that the reliabilities at the output of SISO blocks are referred to binary symbols. This is not always the case, since algorithms like the FB algorithm in the general case outputs reliabilities referring to M-ary symbols, where M is the cardinality of the transmitted information symbol set. However, it is possible to transform M-ary reliabilities into a set of binary reliabilities and vice versa. The two operations are usually non-invertible, thus implying a possible information loss due to the conversion.

As an example, consider the conversion of the probabilities of an M-ary symbol a taking values in the set $\{0, \ldots, M-1\}$ into bit reliabilities. Assume that $M = 2^n$, therefore the symbol a can be represented by n bits $b_0 \ldots b_{n-1}$. Given the probabilities $P\{a = 0\}, \ldots, P\{a = M-1\}$ we compute

$$P\{b_i = 0\} = \sum_{j:b_i=0} P\{a = j\}$$

where $j : b_i = 0$ is the set of all integers $j \in \{0, \ldots, M-1\}$ such that the i-th bit of their binary representation is equal to 0.

The conversion from bit reliabilities to symbol reliabilities can be done as follows, assuming all the bits are independent (which is usually an approximation):

$$P\{a = j\} = \prod_{i=0}^{n-1} P\{b_i = \omega_i(j)\}$$

where $\omega_i(j)$ denotes the i-th bit in the binary representation of j.

As already mentioned, EXIT curves are based on the computation of the MI between each binary symbol and its reliability. Due to the presence of binary symbols, this MI takes on a value between zero and one.

An EXIT curve for a SISO block $\mathbf{S}$ is a function $I_{\mathbf{S}}(I)$ which quantifies the average relationship between the MI of the reliabilities at the input of the block (i.e., the variable I) and the MI of the *a posteriori* reliabilities at the output of the block (i.e., $I_{\mathbf{S}}$)—recall that the MI is computed with respect to the transmitted information sequence [60, 61].

Example 4.2 *Using a Monte Carlo simulation-based method for computing the EXIT curve of a SISO block*

As an example, consider an AWGN inter-symbol interference (ISI) channel with binary phase shift keying (BPSK) at its input. The data is transmitted in blocks of N bits $(a_1, \ldots, a_N)$. At the receiver a SISO block:

- observes the output $\boldsymbol{r}$ of the channel;
- accepts a vector $(m_1^{(\text{in})}, \ldots, m_N^{(\text{in})})$ of N *a priori* probabilities for the transmitted bits, i.e.,
$$m_i^{(\text{in})} = P\{a_i = 1\};$$

- computes a vector $(m_1^{(\mathrm{out})}, \dots, m_N^{(\mathrm{out})})$ of the a posteriori probabilities of the bits using the FB algorithm (2.16), i.e.,

$$m_i^{(\mathrm{out})} = P\{a_i = 1|\boldsymbol{r}\}.$$

Although it has no implication in this particular example, assume also that the output messages represent the extrinsic information, as described in Section 2.4.3. This is a common and important assumption in an iterative detection scheme [46].

Assume that we want evaluate the EXIT curve of the considered SISO block using the previously introduced Monte Carlo simulation-based method.

The SISO block has, as a matter of fact, two (vector) inputs: $\boldsymbol{r}$ and $(m_1^{(\mathrm{in})}, \dots, m_N^{(\mathrm{in})})$. In an iterative decoding process, however, $\boldsymbol{r}$ is fixed, i.e., it does not change during the iterations. The EXIT curve must characterize the MI between the generic transmitted bit a_i and the corresponding output message $m_i^{(\mathrm{out})}$ *as a function of* the MI between the generic transmitted bit a_i and the corresponding input message $m_i^{(\mathrm{in})}$. To this end:

1. fix the SNR at the receiver;

2. generate the bit sequence $(a_1, \dots, a_N)$ using a proper PRNG;

3. simulate the transmission of the bit sequence through the channel, obtaining the vector of observables $\boldsymbol{r}$;

4. generate a vector of messages $(m_1^{(\mathrm{in})}, \dots, m_N^{(\mathrm{in})})$ characterized by an MI equal to I, as will be shortly discussed;

5. run the FB algorithm obtaining the output APPs $(m_1^{(\mathrm{out})}, \dots, m_N^{(\mathrm{out})})$.

Assuming we have a method for generating the input vector $(m_1^{(\mathrm{in})}, \dots, m_N^{(\mathrm{in})})$, the analysis, i.e., the output MI computation, proceeds as follows. Consider the pairs $\{(a_i, m_i^{(\mathrm{out})})\}$ as samples of pairs of RVs distributed according to a pdf $p(a, m)$. Given the sample pair sequence, we wish to estimate the MI between a and m, which represents the output MI. There are several methods for evaluating the MI from a sample sequence. A very simple one is deriving a histogram to estimate $p(a, m)$ and then computing the MI of the corresponding joint discrete RVs. Let us follow this simple method. Note that, in the considered scenario, the messages belong to $[0, 1) \subset \mathbb{R}$ and, therefore, to derive a histogram one needs to define a quantization rule. To this end, divide the interval $[0, 1)$ into L bins $B_1, \dots, B_L$, each of width $1/L$. At this point, it is possible to associate the vector $(m_1^{(\mathrm{out})}, \ldots, m_N^{(\mathrm{out})})$ with a quantized vector $(\tilde{m}_1^{(\mathrm{out})}, \ldots, \tilde{m}_N^{(\mathrm{out})})$, where

$$\begin{aligned}\tilde{m}_i^{(\mathrm{out})} &= \min\{j : m_i^{(\mathrm{out})} \in B_j\} \\ &= \left\lfloor m_i^{(\mathrm{out})} L \right\rfloor .\end{aligned}$$

The correct choice of the number of bins L is important and should be chosen so that $1 \ll L \ll N$. The two extremal choices, $L = 1$ or $L = N$, will result in a MI equal to 0 and 1, respectively. Clearly the quantization operation would not have been necessary, had we analyzed a SISO block operating with quantized messages. The histogram approximation can therefore be obtained based on the sequence $(a_1, \ldots, a_N)$ and the quantized message sequence and is represented by the following joint pmf:

$$\tilde{p}(a, \tilde{m}) = \frac{1}{N} \sum_{i=1}^{N} 1(a_i = a \wedge \tilde{m}_i^{(\mathrm{out})} = \tilde{m})$$

where $1(\cdot)$ is the indicator function previously introduced and $\wedge$ denotes the logical AND.

Given the estimate pmf of the quantized messages, we can compute the MI between a transmitted bit A and the corresponding message M using the following approximation:

$$I(A; M) \simeq \sum_a \sum_{\tilde{m}} \tilde{p}(a, \tilde{m}) \log_2 \frac{\tilde{p}(a, \tilde{m})}{\tilde{p}(a)\tilde{p}(\tilde{m})}$$

where

$$\tilde{p}(\tilde{m}) = \sum_a \tilde{p}(a, \tilde{m})$$

and

$$\tilde{p}(a) = \sum_{\tilde{m}} \tilde{p}(a, \tilde{m}) .$$

By changing the MI I characterizing the input set, and by re-performing all the above described steps, one can obtain a new set of output messages and compute the new output MI $I_S(I) = I(A; M)$, thus obtaining all desired points of the EXIT curve $I_S(I)$.

All the above considerations assume we can generate a vector $(m_1^{(\mathrm{in})}, \ldots, m_N^{(\mathrm{in})})$ of input messages characterized by an MI $I(A; M^{(\mathrm{in})})$ equal to I. A possible solution is starting with an input vector of probabilities all equal to $1/2$, i.e., declaring to the SISO block that there is no *a priori* information on the transmitted bits. This corresponds to start with $I = 0$. We can compute the output message vector and characterize its MI $I_S(0)$. At this point, the output message vector is a vector of probabilities for which we know the MI and, in principle, could be used as input vector for the estimation of $I_S(I_S(0))$. This can be recursively combined to obtain several point of the EXIT curve. However, this approach could lead to inaccurate results and requires to keep the same transmitted bit sequence since all message set will refer to that particular sequence. A common approach to generate an *a priori* probability message sequence for a bit sequence $(a_1, \ldots, a_N)$ is as follows.

Considering a BPSK transmission over an AWGN channel, fix the channel noise variance σ^2 so that the MI between the input and the output of the channel is equal to I. This can be done numerically by inverting the MI expression given in Example 1.1. Transmit the bit sequence $(a_1, \ldots, a_N)$ through the obtained channel, i.e., for each bit a_i apply the BPSK mapping rule, obtaining a transmitted symbol c_i, and add an AWGN noise sample characterized by variance σ^2, obtaining an output observable y_i. Now, compute the *a posteriori* probability of the bit:

$$m_i = \frac{e^{-\frac{(y_i+1)^2}{2\sigma^2}}}{e^{-\frac{(y_i+1)^2}{2\sigma^2}} + e^{-\frac{(y_i-1)^2}{2\sigma^2}}} . \tag{4.6}$$

Since m_i is an invertible function of y_i, by the data processing inequality,

$$I(A; M) = I(A; Y) = I .$$

This implies that the vector $(m_1, \ldots, m_N)$ is a vector of messages, where each element represents the probability that the corresponding transmitted bit is equal to 1 and whose MI (i.e., the MI between the message and the transmitted bit RVs) is equal to I.

Note that there are infinite methods of generating messages characterized by a MI equal to I, and each method is characterized by a conditional distribution of the messages. The above described method is particularly interesting since the generated message sequence, in the log-likelihood domain, is conditionally Gaussian. This is an appealing property, since there are several useful SISO block output messages that are characterized by a Gaussian distribution in the log-likelihood domain. In other words, a message set generated with the above described method will exhibit statistical properties similar to those of a real SISO block.

4.5 EXIT Charts for LDPC Codes

The following example describes a method to compute the EXIT curves associated with the VND and the CND, i.e., the component blocks of a standard LDPC decoder.

Example 4.3 *EXIT curves for the belief propagation LDPC decoder: Gaussian approximation*

Consider the variable and check node algorithm (3.9) and (3.10), involving messages in the LLR domain . Assume to approximate the distribution of the messages with a Gaussian distribution and assume also that all the input messages have equal distribution (which, as stated in the previous section, is a common assumption in density evolution analysis). In [13], it is shown that the distribution $p(m)$ of a message m in an LDPC belief propagation decoder in the LLR domain must fulfill the following symmetry condition:

$$p(m) = e^m p(-m)$$

which, for a Gaussian distribution, implies that

$$\mathrm{Var}\{m\} = 2\mathrm{E}\{m\}\,.$$

Therefore, tracking the variance is sufficient to completely describe the evolution of the distribution.

Assuming, without loss of generality, the transmission of an all-0 sequence, given a Gaussian LLR message set fulfilling the above symmetry condition, the MI between the generic message and the corresponding codeword bit is [69]

$$J(\sigma) \triangleq \int_{-\infty}^{+\infty} \frac{1}{\sqrt{2\pi\sigma^2}} e^{-\frac{(x-\sigma^2/2)^2}{2\sigma^2}} \log_2 \frac{2}{1+e^{-x}}\, dx\,. \tag{4.7}$$

where σ denotes the standard deviation of the message set.

Consider now the operation (3.9) performed during iterative decoding by the variable node and operating in the log likelihood domain. Assuming that the messages at the input of the variable node are independent, the variance of the output message will be equal to the sum of the variances of the input messages. As a consequence, the input-output relation of the MI in a degree d_v variable node, under the Gaussianity assumption, will be:

$$I_{\text{out}} = J\left(\sqrt{(d_v-1)(J^{-1}(I_{\text{in}}))^2 + (J^{-1}(I_0))^2}\right) \tag{4.8}$$

where I_{out} and I_{in} denote the MI between the transmitted codeword bit and a corresponding output and input message, respectively and

1. $$J^{-1}(\cdot)$$

   is the inverse of the $J(\cdot)$ function

2. $$J^{-1}(I_{\text{in}})$$

   is the standard deviation σ_{in} associated with the input message set

3. $$(d_v - 1)(J^{-1}(I_{\text{in}}))^2$$

   is the sum of the variances of the input messages

4. I_0 is the MI between the *external observation* (leading in (3.9) to the LLR m_0) and the corresponding codeword bit

5. $$(J^{-1}(I_0))^2$$

   is the variance of the message associated with the *external observation*

6. and, finally,

   $$\left(\sqrt{(d_v - 1)(J^{-1}(I_{\text{in}}))^2 + (J^{-1}(I_0))^2}\right)$$

   is the standard deviation of the message at the output of the variable node.

If the LDPC code is irregular, one can obtain an average MI associated to the generic message from the VND and input to the CND according to (4.5), which leads to

$$I_{\text{out}}^{\text{VND}} = \sum_i \lambda_i J\left(\sqrt{(i-1)(J^{-1}(I_{\text{in}}))^2 + (J^{-1}(I_0))^2}\right) \tag{4.9}$$

where $\{\lambda_i\}$ are the variable node degree distribution coefficients

An approximate formula for the input/output relation for a degree-d_c check node, based on a property of the BEC, is given by

$$I_{\text{out}} = 1 - J\left(\sqrt{d_c - 1}\, J^{-1}(1 - I_{\text{in}})\right). \tag{4.10}$$

For a detailed overview on how to derive the approximate formula (4.10) and the exact one, we refer the interested reader to [70, 71].

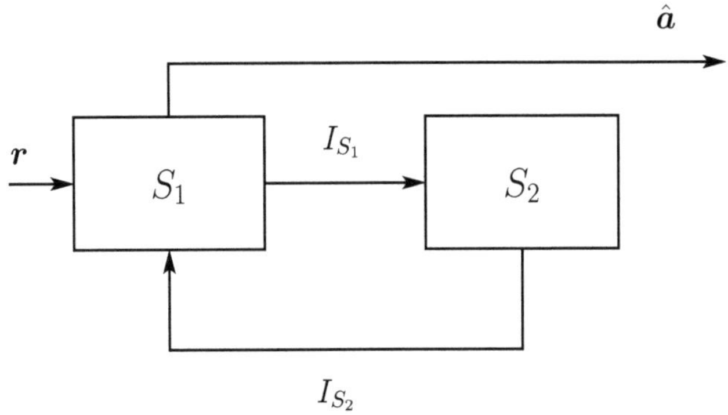

Figure 4.4: Schematic diagram of an iterative receiver comprising two SISO modules.

The average MI at the output of the CND, computed according to (4.5), is as follows:

$$I_{\mathbf{B}} = 1 - \sum_j \rho_j J\Big(\sqrt{j-1}\, J^{-1}(1 - I_{\mathbf{A}})\Big) \tag{4.11}$$

where $\{\rho_j\}$ are the check node degree distribution coefficients.

The MI of the message, computed according to (3.11), at the output of a degree-d_v variable node at the last iteration is

$$I_{\text{out}} = J\Big(\sqrt{(d_v)(J^{-1}(I_{\text{in}}))^2 + (J^{-1}(I_0))^2}\Big) \tag{4.12}$$

and the corresponding average MI is

$$I_{\text{out}}^{\text{VND}} = \sum_i \lambda_i J\Big(\sqrt{(i)(J^{-1}(I_{\text{in}}))^2 + (J^{-1}(I_0))^2}\Big) \,. \tag{4.13}$$

In the following chapters, the receiver will be divided into two distinct processing blocks. This allows to simplify the analysis by decomposing the MI input-output relation into two simpler functions. A generic example of this scheme is shown in Figure 4.4, where two SISO modules S_1 and S_2 are connected in a turbo-like configuration. In Figure 4.5, the EXIT curves of these two hypothetical blocks S_1 and S_2 are shown. In the graph, the horizontal axis refers to the output MI of SISO module S_2 and the vertical axis refers to the output MI of SISO module S_1, i.e., the inverse $I_{S_2}^{-1}(I)$ of the S_2 EXIT curve is actually plotted. This representation of a pair of EXIT curves is referred to as *EXIT chart*, and is useful to investigate the decoding process as a recursive

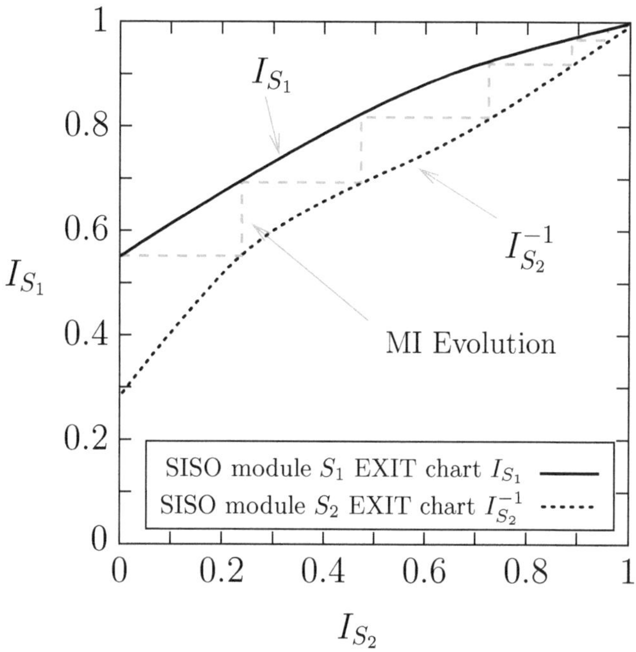

Figure 4.5: Example of EXIT chart: two SISO modules S_1 and S_2 iteratively exchange messages; the evolution trajectory of the MI is also shown.

update of the MI. A trajectory representing the evolution of the MI at the output of the SISO modules S_1 and S_2 in the EXIT chart is also shown. If the MI becomes equal to 1, the decoding process is said to *converge*, in the sense that a low BER can be expected.

In Chapter 5, the relation between the MI and the BER will be investigated and used for LDPC code design purposes.

4.6 Concluding Remarks

In this chapter, the main analysis tools for iterative receivers have been discussed. In particular, EXIT chart-based analyses will play an important role in the rest of the book, since it provides a simplified, yet accurate, convergence analysis tool which is well suited for LDPC code design for coded modulations.

Chapter 5

LDPC Coded Modulations: Analysis and Design

5.1 Introduction

As discussed in the previous chapters, low-density parity-check (LDPC) coding is gaining increasing attention, from an implementation viewpoint, in the scientific community and the industry. In particular, since LDPC codes exhibit near-capacity performance on a variety of memoryless channels, there is high interest in investigating their performance on practical channels.

A simple, although powerful, technique for exploiting LDPC codes over generic channels is based on the use of the *LDPC coded modulations*, which represent the focus of this book. This approach is based on the concatenation of an LDPC encoder and a coded modulator (CM) characterized by "good" properties for transmission over the considered channel. This is the approach followed in [69], where an LDPC encoder is concatenated with a multiple-input multiple-output (MIMO) channel modulator. At the receiver, a soft-input soft-output (SISO) module [66], designed for the considered CM/channel pair, iteratively exchanges soft reliabilities with a standard LDPC decoder. In [69], it is shown how to analyze, using extrinsic information transfer (EXIT) charts [60], the performance of LDPC codes transmitted over a MIMO channel through a proper modulator, with a corresponding soft-input soft-output (SISO) module at the receiver side—a description of a SISO block can be found in Chapters 2 and 4. This system can be interpreted as a special instance of a bit interleaved coded modulation (BICM) scheme with iterative decoding (ID) [72–75]. Usually BICM or BICM-ID would require an interleaver between the encoded bits and the modulator. However if the LDPC is unstructured, i.e., chosen at ran-

DOI: 10.1007/978-3-540-69457-1_5,

dom from a family of LDPC codes defined by their degree distributions, the interleaver is not required. This can be easily seen by observing that an LDPC code followed by a random interleaver is a new LDPC code whose parity check matrix columns are permuted according to the interleaver. This preserves the degree distributions associated with the LDPC code, and more generally, the connection structure of the LDPC code. Therefore, the performance of a belief propagation decoder operating on its graph is also preserved.[1] In practical applications, structured LDPC codes are appealing because the implementations of the corresponding encoder and decoder can take advantage of the structure resulting in lower hardware requirements. In this case, an interleaver might be required after the encoder and its introduction should be carefully considered.

In [69], a heuristic optimization technique for the degree distributions of LDPC codes is also proposed—as a matter of fact, the technique considered in [69] was originally introduced in [76, 77] as a way of designing LDPC codes suited to memoryless channels and approaching the capacity limit.

In this chapter, we first introduce LDPC coded modulations, describing possible detection strategies. Then, we give a detailed description of the general communication system in [69], considering both the transmitter and the receiver sides. The main features of the considered scheme, as compared to a standard BICM-ID scheme, are (i) the possible presence of a modulator *with memory* and (ii) the particular sub-block decomposition of the receiver. We then discuss an EXIT chart-based analysis technique, following the guidelines in [69], and give new insights on the decoding convergence based on the results in [78, 79]. In particular, in order to characterize the extrinsic information evolution on the EXIT chart in terms of bit error rate (BER), we introduce a general upper bound on the BER as a function of the mutual information (MI) between the observables and the transmitted information symbols. These bounds are verified through Monte Carlo simulation-based and density evolution techniques [34]. Finally, we describe a design algorithm for LDPC coded modulations based on the optimization of the degree distributions of the LDPC code [79]. This algorithm is based on EXIT charts and consists of a semi-random walk in the degree distribution parametric space. This optimization algorithm will be applied in a few practical scenarios of interest, with particular attention to inter-symbol interference (ISI) channels

[1]This holds assuming that every pair codeword bit/observable has the same statistical description, i.e., every binary symbol "sees" the same channel. In an LDPC coded modulation, this might not be true and, therefore, specific column permutations might change the system performance. This can be addressed, whenever needed, by a higher order analysis, i.e., an analysis that accounts for different message distributions, depending on the mapping of the codeword bits into symbols.

and partial response channels (PRCs).

5.2 LDPC Coded Modulation Schemes: Basics

In the literature, the design of schemes based on the concatenation of an encoder (either convolutional or block) and a modulator has received substantial attention. In [80, 81], BICM schemes, consisting of the concatenation of a binary encoder, a bit interleaver and a high order memoryless mapper, are proposed and analyzed. At the decoder side, a soft demapper generates reliability values for the bits embedded in each modulated symbol, and these values feed a decoder corresponding to the binary encoder used at the transmitter side. As mentioned in the previous section, in [72–75] an extension of BICM schemes, denoted as BICM-ID, is proposed: iterative information exchange between the soft demapper and the decoder is considered and performance advantages are observed.

As mentioned before, LDPC codes can achieve good performance when transmitted through memoryless channels. However, in many applications, the cascade of modulator, physical channel, and demodulator usually does have memory. As a consequence, it would be interesting to exploit the power of LDPC codes in such more practical scenarios. It should be noted that the near capacity performance of LDPC codes on memoryless channels does not necessarily imply that they can achieve similar performance, i.e., close to the Shannon capacity limit, in other scenarios.

Two main approaches can be devised, from a detection/decoding viewpoint, to allow the use of a standard LDPC code on a generic modulation format and channel. The first approach can be referred to as *graph-based*, whereas the second one can be referred to as *turbo*. These two approaches differ mainly in the receiver design and implementation. In both cases, the transmitter consists of the concatenation of an LDPC encoder and a modulator interfacing with the channel. In the following, we will discuss in detail the assumptions and the implications regarding both approaches

5.2.1 A Graph-Based Detection/Decoding Approach

The corner stone of the graph-based approach has been posed in [82], where the presence of the modulator and the channel, possibly with memory, is taken into account by extending the LDPC code graph to a more general *factor graph* [37] with the introduction of new nodes. These nodes account for the new constraints induced by the modulation and the channel, which increase the correlation between the channel observables beyond the level introduced by the LDPC code [25].

This approach can be seen as an elegant extension of the standard LDPC decoding algorithm. In particular, it becomes appealing if the channel and the modulator can be modeled by simple correlation schemes, as in the case of mapping over a high order constellation, such as M-ary phase shift keying (M-PSK) or quadrature amplitude modulation (QAM) or, for example, transmission of binary PSK (BPSK) over a block fading channel or a phase uncertain channel. The main disadvantage is that the introduction of new constraints in the graph usually leads to the presence of short cycles, which may compromise the performance of the decoding/detection algorithms, if not duly taken into account.

5.2.2 A "Turbo" Detection/Decoding Approach

In this approach, the modulation and the code are treated separately at the receiver. In correspondence to the modulator, at the receiver there is a soft demodulator, or detector, which needs to be designed to take into account the *a priori* probabilities of the modulated codeword bits, for example using techniques similar to those described in Chapter 2 and Section 4.4.[2] Typically, the a priori probability of the transmitted bits is equal to 1/2, i.e., "0" and "1" are equiprobable. However, taking into account generic *a priori* probabilities in the soft demodulator allows to accept information from an external processing block. This block may be an LDPC decoder which exchanges iteratively *extrinsic information* [8, 83] with the soft detector. An illustrative example of this iterative scheme is shown in Figure 5.1, where the soft detector and the soft LDPC decoder are connected in both directions: the soft detector computes messages and sends them to the LDPC decoder which, in turn, computes new messages to be sent to the *a priori* probability input of the soft detector. This technique was originally proposed for turbo decoders [8], from which it borrows the name. In this case, the channel and the modulator are associated with the soft detector and the error correcting encoder is associated with the LDPC decoder. The main disadvantage of this approach is that the subdivision into separate blocks, each accounting for a different aspect of the communication system, is known to be suboptimal. On the other hand, this approach is simple, since it can be based on widely available and well known methods, such as the forward-backward (FB) detection algorithm (see Chapter 2 for more details), the LDPC iterative decoding algo-

[2]The interested reader can find further details in [65, 66].

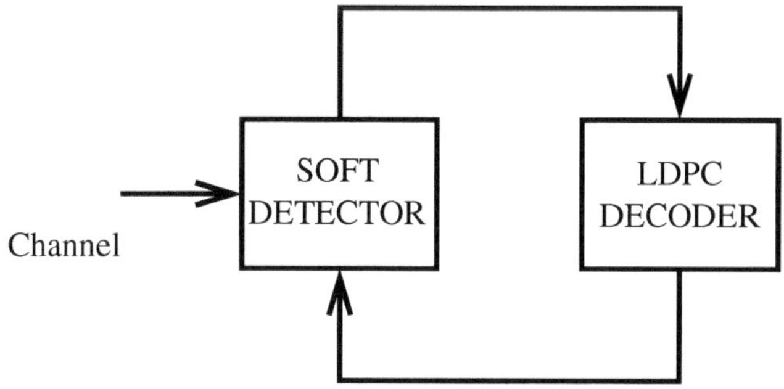

Figure 5.1: Pictorial exemplification of the "turbo approach".

rithm, and communication techniques suitable to combat the specific channel impairments such as, for example, differential encoding (DE), partial response coding etc. Moreover, experience shows that this is a robust approach which allows to devise simple solutions to complex transmission problems still providing remarkable performance. In the remainder of this chapter, this approach will be pursued and its characteristics investigated in more detail.

5.3 Communication System Model

The considered transmitter scheme, shown in Figure 5.2, consists of a simple concatenation of an outer LDPC encoder and an inner CM which is directly connected to the channel. Consider the discrete-time low-pass equivalent model of the communication system. A binary information sequence $\{x_i\}$ at the input of the LDPC encoder is coded into a binary code sequence $\{y_j\}$ (representing a codeword). The binary symbols $\{y_j\}$ are then coded and mapped to high-order modulated symbols $\{c_k\}$. The goal of the inner CM is to make the communication system robust against possible channel impairments. A few possible realistic scenarios include the following:

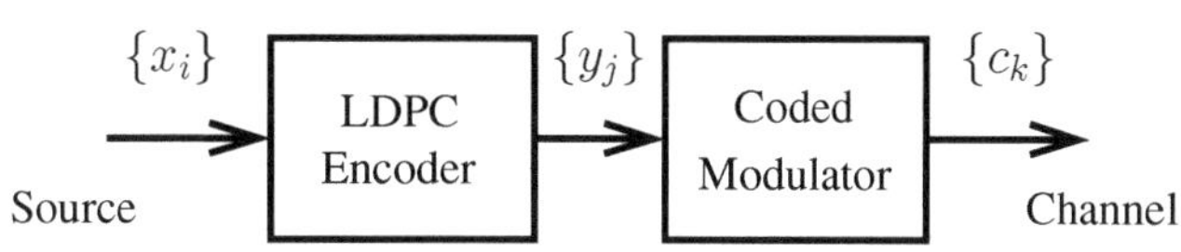

Figure 5.2: System model: transmitter side.

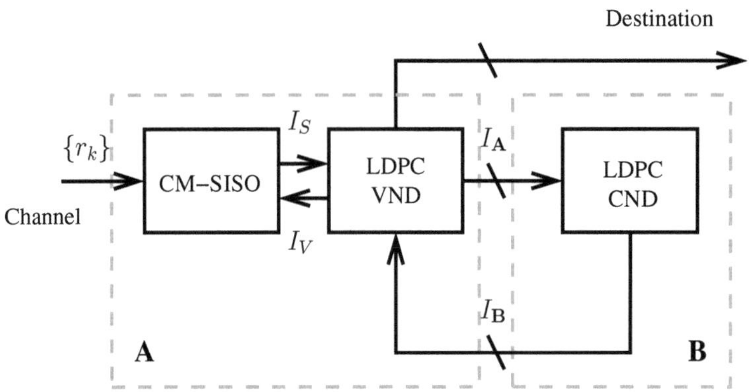

Figure 5.3: System model: receiver side.

- a bandpass communication system where an inner differential encoder is used to solve phase synchronization ambiguities, or to completely avoid phase synchronization problems [84, 85] by means of differential or non-coherent detection [86, 87];
- a system with a modulator which inserts pilot symbols to help synchronization at the receiver side;
- a transmission scheme operating on dispersive or partial response channels;
- a communication system affected by timing uncertainty, where, at the receiver, a SISO block performs soft iterative detection accounting for timing statistical description [88].

In practice, the above scenarios include any scenario comprising a modulator and/or a channel which introduces dependence between the transmitted bits.

The receiver is depicted in Figure 5.3. At the input of the receiver, the sequence of channel observations is denoted as $\{r_k\}$. For simplicity, we are considering one sample per coded symbol. If two or more samples per symbols are necessary in order to cope, for example, with a time varying channel, this derivation can be extended by considering a suitable vector notation.

The receiver is partitioned into two blocks, denoted as **A** and **B**. Block **A** comprises the following sub-blocks.

- A SISO block matched to the CM/channel pair, and referred to as CM-SISO block. This block computes the *a posteriori* reliabilities of the binary symbols $\{y_j\}$ at the input of the CM on the basis of the channel observations and relevant *a priori* reliabilities (coming from the block labeled “LDPC VND” and described below).

- An LDPC Variable Node Detector (VND), associated with the variable nodes in the code bipartite graph. This block computes the reliability of each binary symbol y_j based on the reliabilities from the CM-SISO block and the information received from block **B** and based on the code constraints.

Block **B** includes the LDPC Check Node Detector (CND), associated with the check nodes in the code bipartite graph. The LDPC CND computes the reliability of each binary symbol y_j based on the *a priori* reliabilities received from the LDPC VND and based on the LDPC code constraints.

The reliabilities at the output of block **A**, expressed in the log-likelihood (LL) domain, are computed as follows:

1. the VND processes the messages coming from block **B** by performing, at each variable node, a sum of all the incoming messages *excluding* the one coming from the CM-SISO block; the obtained messages are passed to the CM-SISO block as *a priori* input;

2. the CM-SISO block computes, based on the observations from the channel and the *a priori* information, reliability values according to its internal algorithm (e.g., the FB algorithm [17]);

3. finally, the VND computes the messages to be sent to block **B** according to the standard LDPC decoding algorithm, but using, as *a priori* input, the messages from the CM-SISO decoder.

It is important to observe that, in all the above computations, only the so-called *extrinsic information* is exchanged between the component blocks [8,34].

The overall decoding algorithm at the receiver can be described as follows.

- As initialization step, the *a priori* reliabilities of the symbols $\{y_j\}$ at the input of block **A** (from block **B**) correspond to complete uncertainty (a value equal to 0 in the LL domain).

- Decoding starts from block **A**, which computes output reliabilities and sends them to block **B**. At the first step, since all the messages coming from the CND are 0, the output of block **A** simply consists of the output of the CM-SISO block.

- The CND (i.e., block **B**) thus computes the extrinsic information to be passed to block **A**.
- The algorithm iterates from the second step until a valid LDPC codeword is obtained or a maximum number of iterations N_i is reached.
- If a valid LDPC codeword is not obtained, an additional standard LDPC decoding algorithm is applied based on the last extrinsic information at the input of the VND block. This corresponds to iterating information only between the VND and the CND. The maximum number of standard LDPC decoding iterations is N_{LDPC}.
- At the end of the process, the *complete* (not extrinsic) reliabilities are computed by the VND and delivered to the destination.

The optional N_{LDPC} iterations during which only the VND and CND exchange messages can be exploited in order to reduce the computational burden in all the cases in which a CM-SISO operation is computationally intensive.

5.4 EXIT Chart-Based Performance Analysis

For each block shown in Figure 5.3, it is possible to draw the corresponding EXIT curve [60, 61]. In Figure 5.3, the MI at the output of blocks **A** and **B** is denoted as $I_{\mathbf{A}}$ and $I_{\mathbf{B}}$, respectively. Within block **A**, the MI at the input and output of the CM-SISO sub-block are labeled I_V and I_S, respectively. The decoding process can then be represented as a recursive update of the MI in the EXIT charts. If the MI converges to 1, it is possible to predict that the BER will converge to zero. In fact, let Y be a codeword bit which takes values 1 or 0 with equal probability. Let Z be a reliability referring to Y. As already introduced in Chapter 4, let the MI between Y and Z be

$$I(Y;Z) = H(Y) - H(Y|Z) = 1$$

where $H(\cdot)$ denotes the entropy. Since $I(Y;Z) \geq 0$ and $H(Y) = 1$, it must hold that

$$H(Y|Z) = 0$$

which implies that there exists a function $f_{\mathrm{dec}}(\cdot)$ such that $Y = f_{\mathrm{dec}}(Z)$ with probability equal to 1 [1].

At this point, we are interested in the computation of the EXIT charts of blocks **A** and **B**. Block **B** is simply characterized by the EXIT curve of the LDPC CND, while the EXIT curve of block **A** is obtained by combining the

EXIT curve of the LDPC VND with that of the CM-SISO block. As already noted in Chapter 4 and using the methods proposed in [69], the following approximate formulas for the EXIT curves $I_{\mathbf{A}}$ (of block **A**) and $I_{\mathbf{B}}$ (of block **B**) can be derived:

$$I_{\mathbf{B}} = 1 - \sum_j \rho_j J\Big(\sqrt{j-1}\, J^{-1}(1 - I_{\mathbf{A}})\Big) \tag{5.1}$$

$$I_{\mathbf{A}} = \sum_i \lambda_i J\Big(\sqrt{(i-1)(J^{-1}(I_{\mathbf{B}}))^2 + (J^{-1}(I_S))^2}\Big) \tag{5.2}$$

where the function $J(\cdot)$ is as defined in (4.7), i.e.,

$$J(\sigma) \triangleq \int_{-\infty}^{+\infty} \frac{1}{\sqrt{2\pi\sigma^2}} e^{-\frac{(x-\sigma^2/2)^2}{2\sigma^2}} \log_2 \frac{2}{1+e^{-x}}\, dx\,. \tag{5.3}$$

The MI I_S at the output of the CM-SISO block is a function of the MI I_V of the messages passed by the VND to the CM-SISO block and corresponds to the EXIT function of the CM-SISO block. The MI I_V of the messages passed by the LDPC VND to the CM-SISO block can be approximately computed as follows [69]:

$$I_V = \sum_i \lambda_i J\Big(\sqrt{i} \cdot J^{-1}(I_{\mathbf{B}})\Big)\,. \tag{5.4}$$

As mentioned in Chapter 4, the EXIT curve expressions (5.1), (5.2) and (5.4) are *not* exact. The exact expressions would account for (i) the particular channel, (ii) the particular SISO structure and (iii) the current number of iterations, since all these parameters influence the distribution of the reliabilities which the EXIT chart depends on. Nevertheless, the bounds, although tight, do not accurately predict the actual behavior observed in LDPC CM decoding. We remark that the analysis presented here assumes cycle-free LDPC code graph. This condition can be achieved assuming infinite-length codes. In practice, the absence of cycles of order lower than or equal to 6 is sufficient to guarantee good accuracy of the described analysis.

The "divide and conquer" principle suggests to split the decoding process into phases which can be treated separately. In the following subsections, the iterative decoding is analyzed considering two phases: (1) the first iterations and (2) the asymptotic (with the number of iterations) convergence.

5.4.1 The First Iterations

This is the first part of the decoding process. Consider the illustrative EXIT chart shown in Figure 5.4. The decoding trajectory starts from the intersection

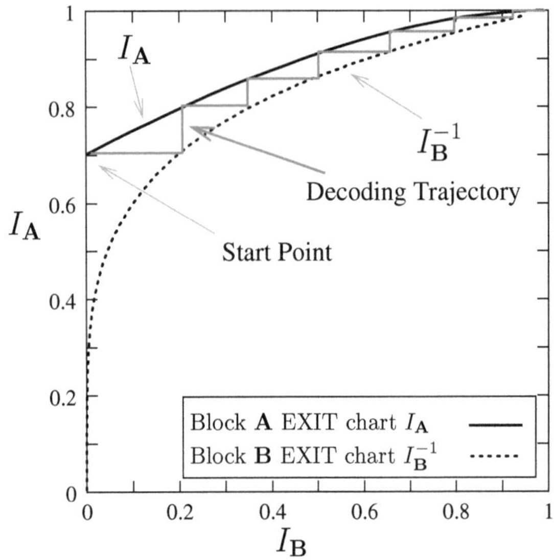

Figure 5.4: Decoding trajectory on an EXIT chart of an LDPC coded modulation scheme.

of the I_A curve with the vertical axis ($I = 0$). As previously discussed, this point represents the MI between the codeword bits and the soft messages at the output of the CM-SISO block when no *a priori* information is available. The decoding process evolves toward the point $(1, 1)$. This decoding phase determines if there will or will not be convergence, i.e., the operating point of the system will fall beyond the BER "waterfall region" (typical of iterative decoding schemes) or before it.

5.4.2 The Asymptotic Convergence Region

Once the MI has reached the neighborhoods of the point $(1, 1)$, any supplementary iteration has the effect of reducing the BER. In this region, the EXIT curves can be approximated based on proper Taylor series expansions and the evolution of the MI towards 1 for increasing number of iterations can be analyzed. This subject will be considered in more detail in the remainder of this chapter.

5.5 Upper and Lower Bounds on the BER

In this section, a relation between the MI and the BER is devised. This will be exploited for code design purposes in the following. Usually, an EXIT chart-based analysis assumes decoding convergence, i.e., sufficiently low BER, when the MI between the codeword bits and the vector of reliabilities has become equal to 1 [60]. This is because, as shown in Section 5.4, a MI equal to 1 implies that the codeword bits can be expressed as a function of the reliabilities with probability one.

Nevertheless, it is usually impossible for the MI to become equal to 1 within a finite number of decoding iterations. On the other hand, if knowledge of the EXIT curves is available, one can compute the evolution of the MI towards 1 as a function of the number of iterations or, given a maximum number of iterations, one can compute the minimum signal-to-noise ratio[3] (SNR) required to achieve a given MI (usually close to 1). Therefore, it becomes useful to have upper and lower bounds on the BER as functions of the MI.

5.5.1 MI-Based Lower Bound on the BER

The reliability values for the codeword bits are usually computed by iterative algorithms in two forms: (i) estimates of the probability of each bit to be equal to 1 and (ii) the corresponding log-likelihood ratios [9]. In general, these reliabilities are random variables (RV) with a given distribution which is a function of several parameters such as, for example, the SNR, the number of iterations, etc. Typically, a reliability value y is used at the end of the iterative processing to make a decision on the corresponding bit x. This has to be done by applying a function $f_{\text{dec}}(\cdot)$ to the reliability value.

We assume binary equiprobable transmission of information bits equal to 1 or 0. We denote by $I = I(X;Y)$ the MI between X (a generic bit) and Y (its corresponding reliability), at given values of the SNR, the number of iterations, etc. A lower bound on the BER, generally denoted as P_{e}, is given by the Fano bound [1]. In fact,

$$
\begin{aligned}
H(X) &= 1 \\
I(X;Y) &= 1 - H(X|Y) \\
H(P_{\text{e}}) &\geq H(X|Y) \qquad (5.5) \\
P_{\text{e}} &\geq H^{-1}[1 - I(X;Y)] \qquad (5.6)
\end{aligned}
$$

[3]Here, the SNR is defined, in general terms, as a parameter which completely defines the channel and such that the MI between the input and the output of the channel is a monotonic function of it.

where inequality (5.5) is the Fano bound applied to a binary RV. The binary entropy function $H(p) \triangleq -p\log(p) - (1-p)\log(1-p)$ is invertible if $p \in (0, 1/2]$. The meaning of inequality (5.6) is that, even choosing the best decision strategy, the BER cannot be lower than $H^{-1}[1-I]$.

5.5.2 MI-Based Upper Bound on the BER

In order to characterize the BER performance, we now derive an upper bound for the BER as a function of the MI. This bound can also be found in [89]. In order to do this, a decision criterion, or, equivalently, a decision function $f_{\mathrm{dec}}(\cdot)$ which maps a reliability to a specific decision bit, has to be chosen. In many cases of practical interest, the reliability of a bit is represented by the *a posteriori probability* for that bit of being "1." Given a set of constraints on the code characteristics, a maximum *a posteriori* (MAP) decision strategy can be implemented in a straightforward manner simply by deciding for the most probable bit. Thus, we investigate MAP decision of the bit given its reliability. In general, this choice requires perfect knowledge of the conditional probability $P\{X = x|Y = y\}$. The following theorem gives an upper bound on the BER, for a given MI $I(X;Y)$, assuming MAP decision.

Theorem 5.1. *Let X be a binary RV such that $P\{X = 0\} = P\{X = 1\} = 1/2$, Y be a RV and $P\{X = x|Y = y\}$ be the conditional pdf of X given Y. Let*

$$\hat{X} = \operatorname*{argmax}_x P\{X = x|Y\}.$$

Given the conditional entropy $H(X|Y) = 1 - I(X;Y)$, then

$$P_{\mathrm{e}} = P\{X \neq \hat{X}\} \leq \frac{H(X|Y)}{2}. \tag{5.7}$$

Proof. The MAP strategy entails a decision for $\hat{x}$, as a function of y, according to the following rule:

$$\hat{x} = \operatorname*{argmax}_x P\{X = x|Y = y\}.$$

This implies that the conditional probability of the error event $E = \{X \neq \hat{x}\}$ given $\{Y = y\}$ can be expressed as

$$\begin{aligned} P\{E|Y = y\} &= 1 - \max_x P\{X = x|Y = y\} \\ &= \min_x P\{X = x|Y = y\} \end{aligned} \tag{5.8}$$

which is, obviously, a number lower than or equal to 1/2. Considering (5.8), one can conclude that

$$H(X|Y=y) = H(P\{X=x|Y=y\}) = H(P\{E|Y=y\})\,. \tag{5.9}$$

Therefore, given $H(X|Y = y)$, assuming that the MAP strategy is used, $P\{E|Y=y\}$ becomes

$$P\{E|Y=y\} = H^{-1}[H(X|Y=y)] \tag{5.10}$$

where $H^{-1}(\cdot)$ is the inverse of the function $H(p)$ for $p \in (0, \frac{1}{2}]$. The following derivation holds:

$$\begin{aligned} P_{\mathrm{e}} &= \mathrm{E}\{P\{E|Y\} \\ &= \mathrm{E}\{H^{-1}[H(X|Y)]\} && (5.11) \\ &\leq \frac{H(X|Y)}{2} && (5.12) \end{aligned}$$

where (5.11) follows from (5.10) and (5.12) follows from the fact that $H^{-1}(x) < x/2, \forall x : 0 \leq x < 1$. □

In Figure 5.5, the lower bound based on the Fano inequality, the upper bound based on MAP decision strategy (Fano lower bound and MAP upper bound, respectively), along with the actual BER performance versus the bit SNR E_b/N_0, are shown considering a regular (3, 6) LDPC code with codeword length 12000. The MI needed for the bounds is computed based on a message distribution obtained using density evolution. The considered numbers of iterations are 3, 6 and 10. The bounds clearly describe the behavior of the BER curve. Thus, the convergence of the MI has useful implications on the convergence of the BER.

5.6 Code Design for LDPC Coded Modulations

In order to design LDPC codes suitable to LDPC coded modulations, we now introduce a code optimization technique based on the use of *EXIT charts* and consisting of a "clever" *random walk* across the parametric space, i.e., the LDPC code degree distributions. The key point for this optimization technique is that, thanks to the underlying EXIT chart-based analysis, it can take into account channel impairments, which are usually neglected for the sake of feasibility by other analysis methods. Moreover, as we will readily see, its simplicity guarantees a low computational complexity.

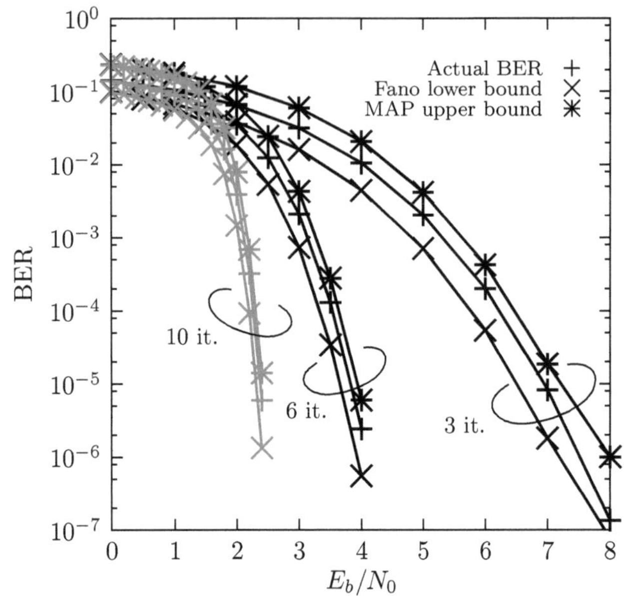

Figure 5.5: Actual BER, MAP upper bound and Fano lower bound for a regular (3, 6) BPSK modulated LDPC code of codeword length 12000 transmitted over an AWGN channel considering 3, 6 and 10 iterations.

5.6.1 The Need for Optimization

A reader might wonder if there is a real need for an LDPC code *optimized specifically* for the used inner CM block, or if the use of a powerful LDPC code designed for an additive white Gaussian noise (AWGN) channel might be sufficient to achieve near-channel capacity performance *regardless* of the inner CM block. In general, the LDPC code has to be optimized for the specific application, and an EXIT chart-based analysis can provide additional insights regarding this aspect.

In Section 5.3, we have chosen to partition the receiver into the two blocks **A** and **B**, including in block **A** both the CM-SISO block and the LDPC VND modules. Note that it would have been possible to study the recursive exchange of information of the three distinct component modules: CM-SISO block, LDPC VND, and LDPC CND. The CM-SISO block and the LDPC CND would have one input and one output (not taking into account the channel input for the CM-SISO module, since the corresponding messages are constant throughout the decoding process) and the LDPC VND would have two inputs and one output. The presence of the feedback from the VND to the CM-SISO is crucial. By removing the feedback the CM-SISO operated as a "one-shot" processing block that elaborates the observed singal from the channel and produces a soft output, i.e., bit reliabilities for the LDPC codeword bits. Such a block could be referred to as a *soft demapper*, without *a priori* input. Clearly, the EXIT curve of block **A** with the CM-SISO module can not be equal to the EXIT curve of block **A** with a soft demapper only. In fact, every time the CM-SISO module input changes, the reliability values to be sent to the variable nodes are recomputed. On the contrary, these reliability values would be unmodified in a scheme with soft demapping only. It follows that, since the EXIT curves of block **A** (interpreted as functions of the parameters $\{\lambda_i\}$) are different in the two cases and since the EXIT curves of block **B** (interpreted as functions of the parameters $\{\rho_j\}$) are equal in the two cases, considering an optimization technique based on EXIT charts, in general an LDPC code optimized in the presence of a CM will be different from an LDPC code optimized in the absence of a CM.

Moreover, one can conclude that LDPC codes optimized for the two scenarios should be equal *if and only if* the EXIT curves of the CM-SISO module are constant and independent of the MI at the feedback input. This applies, for example, to a scenario where the CM-SISO module's feedback input is not used, as in the case of BICM, or for every BI memoryless channel, such as a binary symmetric channel (BSC), a binary erasure channel (BEC), as well as a BI-AWGN channel. These considerations will be useful also in Section 5.7, where examples of "almost flat" CM-SISO EXIT curves will be presented.

5.6.2 Optimizing the EXIT Charts

Consider again the illustrative EXIT chart shown in Figure 5.4: in [69], it is shown that "eye-fitting" the two EXIT curves $I_{\mathbf{A}}(I)$ and $I_{\mathbf{B}}^{-1}(I)$, by varying the degree distributions $(\lambda(x), \rho(x))$, leads to a significant performance improvement. Since the EXIT curves of VND and CND, relative to the most powerful known LDPC codes for memoryless channels, are very similar at "pinch-off," i.e., when EXIT curves touch, and considering the good results obtained in [69], at a first glance fitting the EXIT curves seems a good optimization strategy. However, if only low degree nodes are allowed, the number of degrees of freedom in the fitting procedure becomes small and the best fit might not have good convergence properties. This is particularly important since, in order to construct good codes without short cycles, usually "low degree only" distributions are desirable [9,13]. Moreover, it is important to note that, given a particular SNR (which will be defined exactly later), convergence

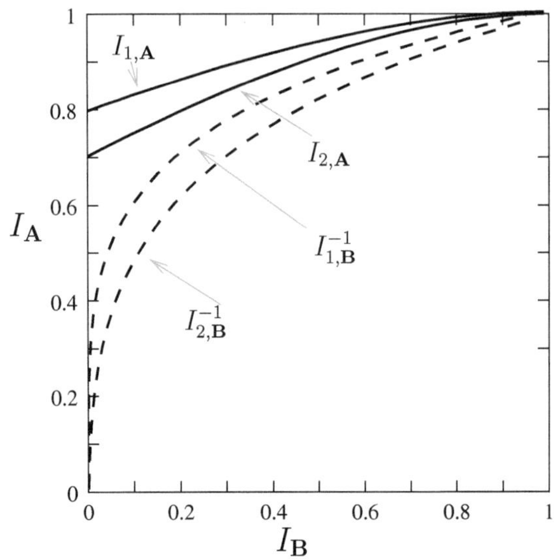

Figure 5.6: Two EXIT curves pairs labeled 1 and 2. Pair 1 has better convergence properties.

of the decoding process can be obtained if the tunnel between the two curves is open. Hence, our actual goal, while performing optimization, *is to keep the tunnel as open as possible*, whereas fitting usually closes the tunnel.

Our optimization algorithm is based on a simple *random walk* in the degree distribution parametric space. Before describing how this algorithm operates, we first provide the reader with some useful considerations and definitions.

Consider Figure 5.6, where two pairs of EXIT curves for blocks **A** and **B**, denoted as $(I_{1,\mathbf{A}}(\cdot), I_{1,\mathbf{B}}^{-1}(\cdot))$ and $(I_{2,\mathbf{A}}(\cdot), I_{2,\mathbf{B}}^{-1}(\cdot))$, respectively, are shown. If

$$\begin{array}{rcll} I_{1,\mathbf{A}}(I) & \geq & I_{2,\mathbf{A}}(I) & \forall I \in (0,1) \\ I_{1,\mathbf{B}}^{-1}(I) & \leq & I_{2,\mathbf{B}}^{-1}(I) & \forall I \in (0,1) \end{array} \tag{5.13}$$

i.e., $I_{1,\mathbf{A}}$ is higher than $I_{2,\mathbf{A}}$ and $I_{1,\mathbf{B}}^{-1}$ is lower than $I_{2,\mathbf{B}}^{-1}$, then the convergence of the decoding process for the system relative to the EXIT curves $(I_{1,\mathbf{A}}(\cdot), I_{1,\mathbf{B}}^{-1}(\cdot))$ will not be slower than the convergence of the system relative to the EXIT curves $(I_{2,\mathbf{A}}(\cdot), I_{2,\mathbf{B}}^{-1}(\cdot))$.

It should be observed that the two EXIT curves touch at the point $(1,1)$— a sufficient condition for this is the absence of degree-1 variable nodes in the code, as it can be seen by imposing $\lambda_1 = 0$ in equation (5.2) and letting $I_{\mathbf{B}} \to 0$. The iterative decoding algorithm for a system characterized by the EXIT curves $(I_{\mathbf{A}}(\cdot), I_{\mathbf{B}}^{-1}(\cdot))$ cannot converge *if* there exists a value $I^\star$, $0 < I^\star < 1$, such that $I_{\mathbf{A}}(I^\star) < I_{\mathbf{B}}^{-1}(I^\star)$, i.e., the tunnel is closed. We then need to define a functional which is representative of the tunnel closure: the more the tunnel is closed, the lower this functional must be. A possible choice is the following:

$$f(\lambda, \rho) = \min_{I \in [0,1]} \left\{ I_{\mathbf{A}}(I) - I_{\mathbf{B}}^{-1}(I) \right\} \tag{5.14}$$

where we have explicitly indicated the dependence of the functional on the degree distributions. Since, as previously observed, the EXIT curves touch at $(1, 1)$, this functional cannot be positive. Moreover, this functional depends also on the particular channel as well as the CM and the CM-SISO block. As previously observed, it is reasonable to assume that increasing the SNR raises the EXIT curve of block $\mathbf{A}$, while decreasing the SNR lowers it. In other words, if the tunnel between the two EXIT curves is at pinch-off, a small SNR increment should be sufficient to open it.

The design parametric space is given by the node degree distributions $\{\rho_j\}$ and $\{\lambda_i\}$. According to the fundamental relations of the degree distributions

$$\begin{aligned} \textstyle\sum_{j=1}^{\infty} \rho_j &= 1 \\ \textstyle\sum_{i=1}^{\infty} \lambda_i &= 1 \end{aligned} \tag{5.15}$$

and

$$\sum_{j=1}^{\infty} \frac{\rho_j}{j} = (1 - R) \sum_{i=1}^{\infty} \frac{\lambda_i}{i}. \tag{5.16}$$

three parameters are linearly dependent on the others. Hence, one has to choose a parameter from the set $\{\lambda_i\}$, a parameter from the set $\{\rho_j\}$, and an additional parameter from either $\{\lambda_i\}$ or $\{\rho_j\}$. The chosen parameters have then to be expressed as functions of the remaining free parameters. There is no constraint on the number of elements of the sets $\{\lambda_i\}$ and $\{\rho_j\}$, provided that these sets (i) are not empty, (ii) contain at least four elements, and (iii) are finite.

We now describe the optimization algorithm. Start with given valid degree distributions associated with a given code rate, according to (5.16), and determined by a tuple of free parameters. If the tunnel is not closed, i.e., $f(\lambda, \rho) = 0$, decrease the SNR until the tunnel closes and $f(\lambda, \rho) < 0$. New tuples of free parameters are then obtained, by repeatedly adding to the previous tuple a Gaussian increment until all inequalities in (5.15) are satisfied. The mean of the Gaussian increment is zero and the standard deviation is used to "tune" the optimization algorithm. From the new tuple, evaluate $\lambda(x)$ and $\rho(x)$ and, consequently, the value $f(\lambda, \rho)$: if this value is larger than the previous one,

Table 5.1: Optimization algorithm: basic steps.

Start	Initialize $\lambda(x)$ and $\rho(x)$ and compute $f(\lambda,\rho)$.
1	While tunnel is open reduce SNR by small steps and compute the final value of $f(\lambda,\rho)$.
2	Find a new (λ',ρ') compatible with code rate at (small) random distance from (λ,ρ).
3	Compute new $f(\lambda',\rho')$; if not larger than previous $f(\lambda,\rho)$ goto step 2, else $(\lambda,\rho) \leftarrow (\lambda',\rho')$.
4	If stop condition is not reached goto 1 else output (λ,ρ) and final SNR.

substitute the previous tuple with the new one. If the tunnel opens, the SNR is decreased again, and the previous steps are repeated. The algorithm stops when a specific requirement is met, such as, for example, the obtained code ensemble corresponds to an EXIT chart with an open (not closed) tunnel for a desired SNR, or a maximum number of steps (in the random walk) is reached. The steps of the optimization algorithm are summarized in Table 5.1. As a possible improvement to the optimization algorithm, one can diminish the step value, i.e., the standard deviation of the Gaussian increment vector, after a given number of unsuccessful trials. Unlike the EXIT curve fitting optimization algorithm in [69, 90], the described optimization technique offers the advantage of being effective also for small sets of possible node degrees. This algorithm can also be seen as a particular instance of the so called "differential evolution" algorithm [91].

The described algorithm basically performs an optimization of the convergence threshold, defined as the lowest SNR such that the tunnel is open. Within the approximation of the EXIT chart-based analysis, the decoding process converges above this SNR threshold. The simplicity of the proposed optimization algorithm allows a joint optimization of both degree distributions $\{\lambda_i\}$ and $\{\rho_j\}$ in the presence of the CM-SISO block. This would be difficult to perform using analytical optimization techniques. We also observe that, although the predicted thresholds are not very accurate in an absolute sense, they turn out to be proportional to the experimental thresholds obtained by Monte Carlo simulations for actual codes chosen according to optimized degree distributions. This observation, together with the good results shown in the next sections, confirms the validity of the proposed optimization algorithm.

5.7 Code Design for Fast Decoding Convergence

In the previous sections, the convergence of the decoding process for LDPC coded modulations has been investigated in a "boolean sense," meaning that the convergence has been treated as a condition which either can or cannot occur. This allows the design of LDPC codes guaranteeing convergence, i.e., near error free performance, above the threshold SNR.

In this section, we provide deeper insights into the decoding convergence of LDPC coded modulations. The adopted receiver scheme is that considered in Section 5.3, with its block decomposition. We characterize the behavior of the EXIT curves in the neighborhood of the point $(1, 1)$ in the EXIT chart, i.e., the point of successful decoding convergence. We highlight the dependence of the EXIT curves of the decoding blocks on the LDPC code parameters, i.e., the degree distributions, paying particular attention to the *final iterations* needed for convergence. A bound on the LDPC code parameters, first introduced in [78], is given as a necessary condition for decoding convergence. This bound is used in order to justify some of the results obtained in the previous chapters regarding the structure of LDPC codes optimized for specific coded modulations and channels, such as DE-PSK on AWGN channel and ISI channels [79,92]. Moreover, we use the BER bounds, functions of the MI, obtained in Section 5.5 and describe a code design criterion based on mixed MI-BER behaviour as a function of the number of decoding iterations [78].

As an illustrative application of the MI-BER evolution criterion we investigate LDPC codes optimized for a BEC. We fix the number of decoding iterations and optimize the descriptive parameters of the LDPC code, i.e., the degree distributions [13], in order to achieve a given target BER with the "worst possible channel." The obtained code structures are significantly different from those obtained optimizing for convergence threshold. In particular, if a small number of iterations (i.e., between 4 and 8) and a low target BER (i.e., lower than 10^{-3}) are considered, the best codes are very similar to *regular*

LDPC codes whose variable node degree is a monotonic function of the target BER. The described LDPC code optimization framework carried out for a simple BEC can be straightforwardly applied to several transmission schemes with arbitrary modulations and channels, both with and without memory.

Consider the iterative receiver for LDPC codes concatenated with a modulator designed to cope with the specific transmission channel described in Section 5.4. As shown in Figure 5.3, the receiver can be decomposed into two main blocks: (i) block **A** comprising a SISO module for the modulator and the set of all variable nodes, denoted as VND and (ii) block **B** comprising the set of all check nodes, denoted as CND.

Since blocks **A** and **B** exchange iteratively vectors of real valued reliabilities associated with the transmitted LDPC codeword and since for each block these vectors can be computed as a function of the observed received signal and of the vector incoming from the other block, an analysis of the decoding convergence based on EXIT charts can be performed [60, 69]. The analysis based on EXIT charts tracks the evolution of the MI between the codeword bits and their corresponding reliabilities. As discussed in Section 5.4, this analysis is based on the assumption that the MI associated with the output reliability vector of a block is a function only of the MI associated with the input vector (and a function of the channel statistical description as well). As discussed in Section 5.4, this assumption represents an approximation which, in practical situations, turns out to be quite accurate in predicting the system performance [60].

In Figure 5.4, the decoding trajectory of the MI on a generic EXIT chart is shown, with reference to the decoding scheme in [69]. The upper curve is the EXIT curve of block **A**, which comprises the CM-SISO block and the VND. The lower curve is the inverse of the EXIT curve of block **B**, which comprises the CND. It is easily recognized that in order for the MI to converge to 1, the EXIT curve of block **A**, i.e., $I_{\mathbf{A}}(I)$, and the inverse of the EXIT curve of block **B**, i.e., $I_{\mathbf{B}}^{-1}(I)$, must satisfy the following condition:

$$I_{\mathbf{A}}(I) > I_{\mathbf{B}}^{-1}(I) \qquad 0 < I < 1 \,. \tag{5.17}$$

Moreover, given the behavior of $I_{\mathbf{A}}(I)$ and $I_{\mathbf{B}}^{-1}(I)$ in proximity of the point $(1, 1)$, one can completely characterize the convergence of the MI as a function of the number of iterations.

In Section 3.3, the degree distributions of an LDPC code were defined as a couple of polynomials $\lambda(x) = \sum_i \lambda_i x^{i-1}$ and $\rho(x) = \sum_j \rho_j x^{j-1}$, where the coefficients $\{\lambda_i\}$ and $\{\rho_j\}$ denote the fraction of edges in the LDPC code graph connected to degree-i variable nodes and degree-j check nodes, respectively.

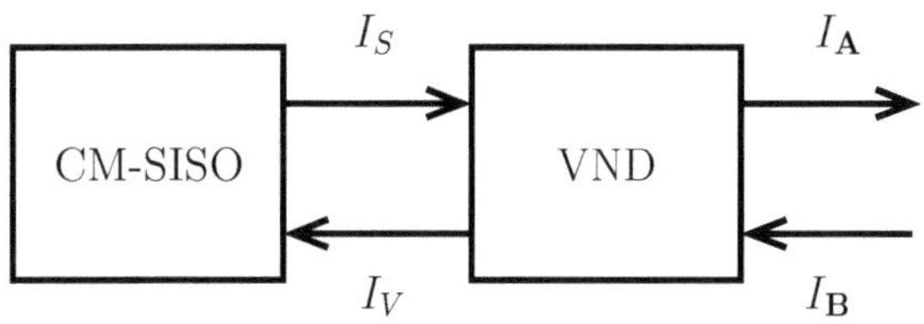

Figure 5.7: Illustrative representation of block **A**.

In Figure 5.7, an illustrative representation of block **A** is given and the MI related to the reliability vectors involved in the computation is shown. In particular, the functional relationships of the MI values can be devised: $I_{\mathbf{A}}$ is a function of $I_{\mathbf{B}}$ and I_S; I_S is a function of I_V, which is also a function of $I_{\mathbf{B}}$. Therefore, one can write

$$I_{\mathbf{A}} = I_{\mathbf{A}}(I_{\mathbf{B}}, I_S(I_V(I_{\mathbf{B}}))).$$

In order to characterize the decoding convergence, we now compute the first order Taylor series expansion of $I_{\mathbf{A}}$ as a function of $I_{\mathbf{B}}$ centered at $I_{\mathbf{B}} = 1$, i.e., in the proximity of error free performance. The derivative of $I_{\mathbf{A}}$ is as follows:

$$\frac{\mathrm{d}I_{\mathbf{A}}}{\mathrm{d}I_{\mathbf{B}}} = \frac{\partial I_{\mathbf{A}}}{\partial I_{\mathbf{B}}} + \frac{\partial I_{\mathbf{A}}}{\partial I_S}\frac{\partial I_S}{\partial I_V}\frac{\partial I_V}{\partial I_{\mathbf{B}}}. \tag{5.18}$$

On the other hand, $I_{\mathbf{A}}$ can be expressed as a linear combination of terms depending on $I_{\mathbf{B}}$ and I_S, weighed by the coefficients $\{\lambda_i\}$ [69, 77]:

$$I_{\mathbf{A}} = \sum_i \lambda_i I_V^{(i)}(I_{\mathbf{B}}, I_S) \tag{5.19}$$

where $I_V^{(i)}$ is the EXIT function associated with a generic degree-i variable node. In the process of obtaining the reliabilities associated to I_V, at the input of the SISO block corresponding to the channel/coded modulator pair (CM-SISO), a generic degree-i variable node acts as a degree-$(i+1)$ variable node with no *a priori* information, i.e., with input *a priori* probability equal to 1/2. This is due to the symmetry in the computation of variable nodes with respect to the messages coming from both the graph and the channel, i.e., the *a priori* probability [9]. As a consequence,

$$I_V = \sum_k \lambda_k I_V^{(k+1)}(I_{\mathbf{B}}, 0). \tag{5.20}$$

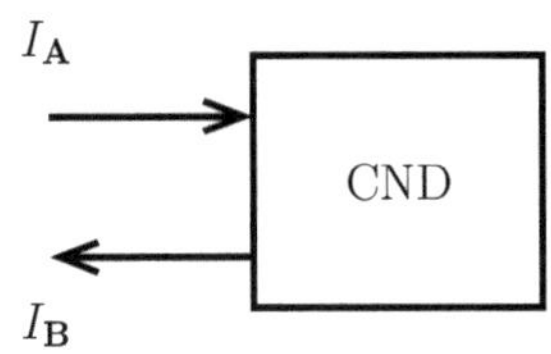

Figure 5.8: Illustrative representation of block **B**.

Substituting (5.19) and (5.20) into (5.18), one obtains

$$\frac{\mathrm{d}I_{\mathbf{A}}}{\mathrm{d}I_{\mathbf{B}}} = \sum_i \lambda_i \left[\frac{\partial I_V^{(i)}}{\partial I_{\mathbf{B}}} + \frac{\partial I_V^{(i)}}{\partial I_S} \frac{\partial I_S}{\partial I_V} \sum_k \lambda_k \frac{\partial I_V^{(k+1)}}{\partial I_{\mathbf{B}}} \right] \tag{5.21}$$

where $\frac{\partial I_S}{\partial I_V}$ denotes the derivative of the CM-SISO EXIT curve with respect to the MI I_V as its *a priori* input.

The block **B**, as shown in Figure 5.8, comprises the CND. Similarly to what has been done for block **A**, the EXIT curve of block **B** can be expressed as a weighed linear combination of the degree distribution $\{\rho_j\}$ as follows [69, 77]:

$$I_{\mathbf{B}} = \sum_j \rho_j I_C^{(j)}(I_{\mathbf{A}}) \tag{5.22}$$

where $I_C^{(j)}$ denotes the EXIT function associated with a generic degree-j check node. From (5.22), one obtains that the derivative of $I_{\mathbf{B}}$, with respect to $I_{\mathbf{A}}$, can be written as

$$\frac{\mathrm{d}I_{\mathbf{B}}}{\mathrm{d}I_{\mathbf{A}}} = \sum_j \rho_j \frac{\mathrm{d}I_C^{(j)}}{\mathrm{d}I_{\mathbf{A}}}(I_{\mathbf{A}}). \tag{5.23}$$

In [93], it has been shown that the reliability distributions in the convergence region, i.e., around to the point $(1,1)$ in the EXIT charts, due to a large variance of the reliability values, tend to behave like the distributions found for a BEC. In other words, the reliabilities tend to group into high valued ones, i.e., "sure" decisions, and low valued ones, i.e., "erasure-like." This fact suggests that in the convergence region one can approximate the information transfer functions $I_V^{(i)}(\cdot,\cdot)$ and $I_C^{(j)}(\cdot)$ with the BEC EXIT functions for a single parity check node and a single variable node [63, 94]:

$$\begin{aligned} I_V^{(i)}(I_{\mathbf{B}}, I_S) &\simeq 1 - (1 - I_S)(1 - I_{\mathbf{B}})^{i-1} && (5.24) \\ I_C^{(j)}(I_{\mathbf{A}}) &\simeq I_{\mathbf{A}}^{j-1}. && (5.25) \end{aligned}$$

The particular cases corresponding to $I_C^{(2)}$, $I_V^{(1)}(I_{\mathbf{B}}, I_S)$, and $I_V^{(2)}(I_{\mathbf{B}}, 0)$ need no approximation, since in these three particular cases the check and variable nodes act as identity blocks:

$$\begin{aligned} I_C^{(2)}(I_{\mathbf{A}}) &= I_{\mathbf{A}} \\ I_V^{(1)}(I_{\mathbf{B}}, I_S) &= I_S \\ I_V^{(2)}(I_{\mathbf{B}}, 0) &= I_{\mathbf{B}} . \end{aligned} \tag{5.26}$$

We remark that from (5.19), (5.22), (5.25), and (5.27), one can derive the following facts:

- $I_{\mathbf{B}}(1) = 1$;
- $I_{\mathbf{A}}(1, I_S) = 1 - \lambda_1 + I_S \lambda_1$. In particular, $I_{\mathbf{A}} < 1$ if $\lambda_1 > 0$ and $I_S < 1$ for $I_{\mathbf{B}} = 1$.

In other words, since $I_{\mathbf{B}}(1) = 1$, in order for the decoding process to converge it must hold that either $\lambda_1 = 0$ or $I_S = 1$ for $I_{\mathbf{B}} = 1$.

We now substitute the previously given approximations (5.24) and (5.25) into (5.21) and (5.23) in order to find the first order Taylor series approximation of the EXIT curves of blocks **A** and **B** at the point $(1, 1)$:

$$\frac{\partial I_V^{(i)}}{\partial I_{\mathbf{B}}}(1, I_S) = \begin{cases} 0 & i = 1 \\ 1 - I_S & i = 2 \\ 0 & \text{otherwise} \end{cases}$$

$$\frac{\partial I_V^{(i)}}{\partial I_S}(1, I_S) = \begin{cases} 1 & i = 1 \\ 0 & \text{otherwise.} \end{cases}$$

Substituting these relations into (5.21), one obtains

$$\frac{\mathrm{d}I_{\mathbf{A}}}{\mathrm{d}I_{\mathbf{B}}}(1) = \lambda_1^2 \frac{\partial I_S}{\partial I_V} + \lambda_2 (1 - I_S) \, . \tag{5.27}$$

Moreover, from (5.23) and (5.25) it follows that

$$\frac{\partial I_{\mathbf{B}}}{\partial I_{\mathbf{A}}}(1) \simeq \sum_j (j-1)\rho_j \, .$$

The derivative of the inverse of the EXIT curve of block **B** is therefore given by

$$\frac{\partial I_{\mathbf{B}}^{-1}}{\partial I_{\mathbf{A}}}(1) = \frac{1}{\sum_j (j-1)\rho_j} \, .$$

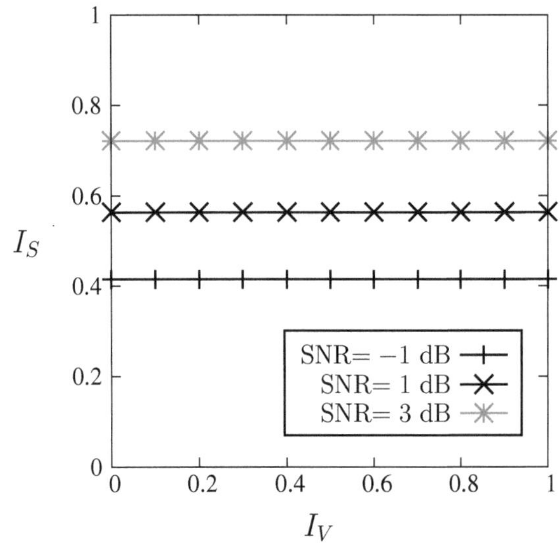

Figure 5.9: EXIT curve I_S, as a function of I_V, of the CM-SISO block for BPSK and AWGN channel.

In order for the decoding process to converge, (5.17) must hold and the derivative of the two functions must satisfy the following inequality:

$$\lambda_1^2 \frac{\partial I_S}{\partial I_V} + \lambda_2(1 - I_S) < \frac{1}{\sum_j (j-1)\rho_j}. \tag{5.28}$$

This bound gives a relation between λ_1, λ_2, $I_S(I_V)$ and $\{\rho_j\}$, which represents a necessary condition that an LDPC coded modulation system must satisfy in order to reach decoding convergence.

In the following, some examples of applications of the results above are given.

Example 5.1 Consider a single LDPC coded communication system with BPSK transmission over an AWGN channel. In this case, the CM block is simply the BPSK modulator and the CM-SISO block could consist of a block performing a symbol-by-symbol conversion from the received sample domain to the log-likelihood *a priori* probability domain. Since no side information is needed by the CM-SISO block to perform this task, the associated EXIT curve I_S is a constant function of the MI I_V of the reliabilities passed by the VND to the CM-SISO block. This is shown explicitly in Fig 5.9, obtained through computer simulations. Since $I_S(1) < 1$, it must hold that $\lambda_1 = 0$.

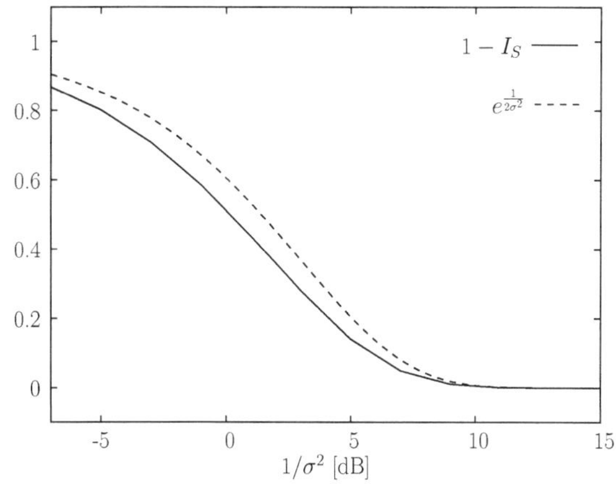

Figure 5.10: Comparison of two factors in the bounds on λ_2 for BPSK and AWGN channel.

Moreover, the bound (5.28) becomes

$$\lambda_2 < \frac{1}{(1 - I_S)\sum_j (j-1)\rho_j}. \tag{5.29}$$

This condition can be directly related to the following stability condition given in [13]:

$$\lambda_2 < \frac{e^{\frac{1}{2\sigma^2}}}{\sum_j (j-1)\rho_j} \tag{5.30}$$

where σ^2 is the variance of the additive noise sample and $1/\sigma^2$ is the SNR. In (5.29) the factor $1 - I_S$ plays the role of the factor $e^{-\frac{1}{2\sigma^2}}$ in (5.30). The two factors are shown in Figure 5.10. Clearly the two values are close, thus confirming the validity of the bound obtained using the EXIT chart analysis.

Example 5.2 In Chapter 7, we will consider code optimization for DE M-ary PSK (MPSK) LDPC coded modulations. The optimized LDPC codes show a structure very different from that of standard LDPC codes for the AWGN channel. In particular, the fraction of degree-2 variable node λ_2 is significantly increased. In Figure 5.11, EXIT curves for a DE-QPSK CM-SISO block are shown, for various values of the SNR. One can notice that $I_S(1) = 1$. Observing that $(1 - I_S) \simeq 0$ in the denominator of (5.29), it follows that the bound on λ_2 is relaxed, thus allowing a larger optimized value for this

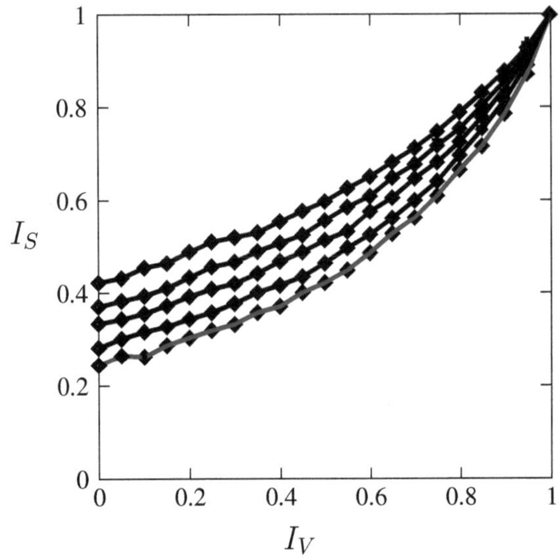

Figure 5.11: EXIT curves I_S, as a function of I_V, of the CM-SISO block for DE-QPSK and AWGN channel. The various curves correspond to different values (equally spaced by 0.5 dB) of SNR, from 0 dB (bottom curve) to 2 dB (top curve).

coefficient. In Chapter 7, LDPC codes optimized for differential modulation will be shown to have, in fact, degree distributions characterized by a large value of λ_2. Since the optimization algorithm adopted in Chapter 7 optimizes the global convergence threshold, it is very likely that the increased value of λ_2 allows to achieve a better decoding threshold at the expense of decoding convergence speed.

Example 5.3 In [92,95], examples of optimized LDPC codes are given both for an AWGN channel affected by ISI and for a partial response channel (PRC). In particular, optimized codes in these scenarios may significantly differ from AWGN LDPC codes if the channel impulse response is long enough. Section 5.9.1 will be devoted to the design of LDPC coded modulations for ISI channels and PRCs. However, we now present a few results, relative to ISI channels, in terms of degree distributions.

In Figure 5.12, the EXIT curves of CM-SISO blocks for a few ISI channels are shown. In particular, the SNR is fixed and ISI channels with impulse

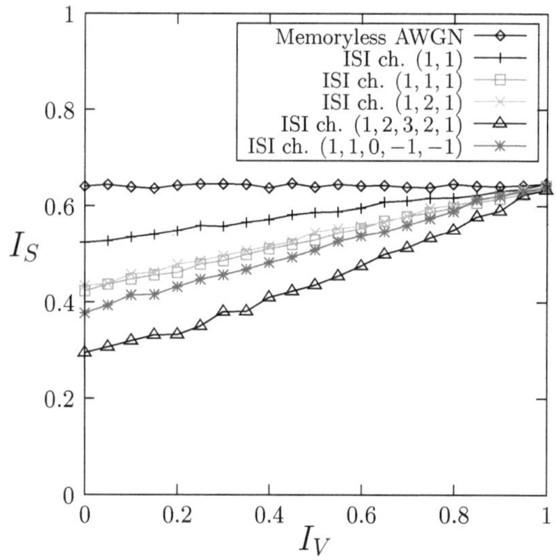

Figure 5.12: EXIT chart I_S of the CM-SISO block for QPSK and several ISI channels.

response coefficients proportional to[4] $(1,1)$, $(1,1,1)$, $(1,2,1)$, $(1,2,3,2,1)$ and $(1,1,0,-1,-1)$ are considered. As noted in [92], a linear trend of the EXIT curves is easily recognized, as well as an increase in the slope of the EXIT curves for longer channel impulse response lengths. Since $I_S(1) < 1$, degree-1 variable nodes are not allowed. The bound on λ_2 is equal to that for BPSK transmission over the AWGN channel. Nevertheless, due to the reduced QPSK input channel capacity of the ISI channels, the SNR needed to achieve convergence at a given code rate is higher that that needed for the AWGN channel. This leads to a larger value of $I_S(1)$ and, therefore, to a larger allowed value for λ_2, bounded as follows:

$$\lambda_2 < \frac{1}{(1 - I_S)\sum_j (j-1)\rho_j}.$$

We verified this result by optimizing LDPC codes for ISI channels. The resulting codes exhibit high values for λ_2 in the case of channels with long impulse response [92].

[4]The energies of the considered ISI channel impulse responses are normalized to 1.

5.8 Code Design for a Target BER

The results described in the previous section suggest to exploit the knowledge of EXIT functions in the proximity of the point $(1, 1)$ of the EXIT charts to analyze (and determine) the convergence of MI (see Section 5.7). Since bounds are available which "link" the MI to the BER (see Section 5.5), it is possible to design LDPC codes in order to achieve a target BER after a fixed number of decoding iterations.

The considered analysis does not take into account *cycles* in the code graph [9, 34]. Cycles are related with three important LDPC coded system parameters: (i) the codeword length, (ii) the maximum number of iterations, and (iii) the maximum allowed node degree. In general, the decoding process delivers a performance close to that of maximum *a posteriori* probability decoding if: (i) the codeword length is large, (ii) the number of iterations needed for convergence is kept small, and (iii) the maximum node degree is low.

On the basis of considerations similar to the previous ones, in [78] an LDPC code design method was proposed based on the convergence of the MI to a given value in a given number of iterations. If the number of iterations required for convergence is small, the possible presence of short cycles should not impact the performance of the system, since these reliabilities whose accuracy is affected by the short cycles cannot propagate through the graph. This enables the use of short codes, i.e., with short codeword lengths, which are difficult to design without short cycles in the graph. In the following, we provide an applicative example of the above described criterion.

Given a system such that both the derivatives, evaluated at 1, of the EXIT functions of blocks **A** and **B** are known and non-zero, the convergence law can be derived as follows.

In the neighborhood of $(1, 1)$ the EXIT curves $I_{\mathbf{A}}$ and $I_{\mathbf{B}}$ can be approximated by their first order Taylor series expansions:

$$\begin{aligned} I_{\mathbf{A}}(I) &\simeq 1 - a(1 - I) \\ I_{\mathbf{B}}(I) &\simeq 1 - b(1 - I) \end{aligned}$$

where

$$\begin{aligned} a &= \lambda_1^2 \frac{\partial I_S}{\partial I_V} + \lambda_2(1 - I_S(1)) \\ b &= \sum_j (j - 1)\rho_j . \end{aligned}$$

The recursion characterizing the decoding behavior is

$$\begin{aligned} I_{2n+1} &= 1 - a(1 - I_{2n}) \\ I_{2n+2} &= 1 - b(1 - I_{2n+1}). \end{aligned} \tag{5.31}$$

Substituting the variable $H_n = 1 - I_n$ in the recursion (5.31), one obtains the following recursion:

$$\begin{aligned} H_{2n+1} &= aH_{2n} \\ H_{2n+2} &= bH_{2n+1}. \end{aligned} \tag{5.32}$$

The start point of the recursion (5.31) is I_{2n_0} (or H_{2n_0} for recursion (5.32)) where n_0 is the number of iterations needed to reach the convergence region, i.e., the region in which the first order Taylor series approximation for $I_{\mathbf{A}}$ and $I_{\mathbf{B}}$ holds.

Solving (5.32) gives

$$H_{2(n+n_0)} = (ab)^n H_{2n_0}$$

and

$$I_{2(n+n_0)} = 1 - (ab)^n(1 - I_{2n_0})$$

which is the MI at the $(n_0 + n)$-th iteration. Applying the bound (5.7) in Theorem 5.1, one obtains:

$$\mathrm{BER}_{n+n_0} \leq (ab)^n \frac{H_{n_0}}{2} \tag{5.33}$$

where BER_n denotes the BER after n iterations. Inequality (5.33) allows the computation of the minimum number of additional iterations to be performed, starting from n_0, in order to obtain the desired BER:

$$n_{\min} = \frac{\log(2\,\mathrm{BER}/H_{n_0})}{\log(ab)}.$$

An alternative, more useful, approach could be based on the design of the degree distributions. Towards this end, by simple manipulations of (5.33) one obtains

$$ab \leq \left(\frac{2\mathrm{BER}}{H_{n_0}}\right)^{1/n}$$

and, therefore,

$$\left[\lambda_1^2 \frac{\partial I_S}{\partial I_V} + \lambda_2(1 - I_S(1))\right] \sum_j (j-1)\rho_j \leq \left(\frac{2\mathrm{BER}}{H_{n_0}}\right)^{1/n}$$

which represents a design constraint guaranteeing convergence to the desired BER in $n + n_0$ iterations.

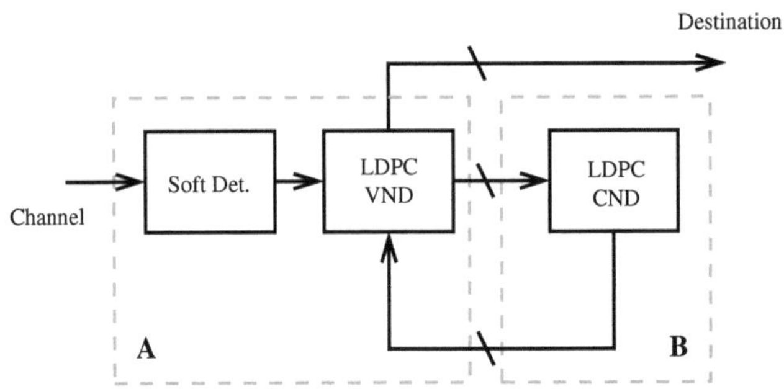

Figure 5.13: Structure of a standard belief propagation receiver for LDPC coded modulations and binary input memoryless channel.

5.8.1 Target BER-based LDPC Code Design for BEC

We now investigate the structure of LDPC codes designed for convergence to a specified BER after a given number of iterations, considering a BEC. A modified version of the design algorithm presented in Section 5.6 is used. By optimizing LDPC codes for a wide variety of conditions, we obtain insights on the structure of LDPC codes suited for a small number of iterations. A discussion on similar code design conditions can be found in [96], where an LDPC code optimization algorithm is proposed for a BEC and the codes are optimized in order to minimize the number of iterations needed to achieve a target BER.

In Figure 5.13, the structure of a belief propagation iterative receiver for LDPC coded modulations is depicted. As usual, the receiver is divided into two main blocks, exchanging vector reliabilities: the block labeled **A**, comprising a soft detector for the channel output and the VND, is associated with the LDPC code variable nodes; the block labeled **B**, comprising the CND, is associated with the set of LDPC code check nodes. The main difference with respect to the system presented in Section 5.3 is the absence of feedback from the VND to the soft detector. This is due to the fact that the BEC is a memoryless binary channel and, therefore, does not take any advantage from *a priori* feedback.

We use the MI between the transmitted codeword bits and the exchanged reliabilities as a measure of the overall achieved system reliability, i.e., we perform an EXIT chart-based analysis. As discussed in Section 5.4, the EXIT chart-based analysis assumes that the MI at the output of each block in the detector's scheme is a single-valued function of the MI of the reliabilities at the input of the block.

In Section 5.5, it is shown that the achievable BER after a given number of iterations can be bounded as follows:

$$H^{-1}[1 - I(X;Y)] \leq \text{BER} \leq \frac{1 - I(X;Y)}{2} \tag{5.34}$$

where $I(X;Y)$ is the achieved mutual information between the generic information bit X and the associated reliability Y. The bounds in (5.34) are useful, since they allow an accurate estimate of the BER obtained by a system when the MI between a generic coded bit and the set of the reliability messages associated with this bit is close to 1. In the following, we will use the upper bound in (5.34) as an estimate of the BER after a given number of LDPC decoder iterations. As previously proposed, the BER will be used as a functional which has to be minimized with respect to the descriptive parameters of the LDPC code, i.e., the degree distributions.

We now characterize the relationship between the number of iterations and the BER, when the iterative LDPC decoder is in the convergence region, i.e., in the last few iterations needed to achieve the desired BER. We focus on a BEC due to its simplicity. Nevertheless, we remark that the characterization relative to a BEC is useful for other channels as well, since it is known that when the decoding process converges at low BER, the system behavior approaches that of an iterative decoder operating on a BEC [93].

The EXIT functions of the VND and CND are given by

$$I_V(I) = 1 - \sum_i \lambda_i (1 - I_{\text{ch}})(1 - I)^{i-1} \tag{5.35}$$

where I_{ch} is the BEC channel capacity, and

$$I_{\mathbf{B}}(I) = \sum_j \rho_j I^{j-1} \tag{5.36}$$

respectively [63, 94]. The decoding process converges if the recursion

$$\begin{array}{rcll} I_0 & = & I_{\text{ch}} & \\ I_n & = & I_{\text{v}}(I_{\mathbf{B}}(I_{n-1})) & \quad n \geq 1 \end{array} \tag{5.37}$$

tends to 1 as n tends to infinity. Approximating (5.36) with its first-order Taylor series expansion at the point 1, one obtains

$$I_{\mathbf{B}}(I) \simeq 1 - (1 - I) \sum_j \rho_j (j - 1) \,. \tag{5.38}$$

One can also approximate $I_V(I)$ by a Taylor approximation at 1, choosing the minimum order such that the resulting function is not constant:

$$I_V(I) \simeq 1 - \lambda_{i_{\min}} (1 - I_{\text{ch}})(1 - I)^{i_{\min}-1} \tag{5.39}$$

where $i_{\min}$ is the lowest variable node degree.

Suppose now that after n_0 iterations the MI is sufficiently close to 1 such that (5.38) and (5.39) are good approximations. After n supplementary iterations, the MI is given by the following relation, obtained by substituting (5.39) and (5.38) into (5.37) and writing explicitly the recursion:

$$\begin{aligned} I_{n+n_0} \quad \simeq \quad & 1 - \left[(1 - I_{\text{ch.}})\lambda_{i_{\min}}\right]^{\frac{b^n-1}{b-1}} \\ & \cdot \left[\sum_j \rho_j (j-1)\right]^{b\frac{b^n-1}{b-1}} (1 - I_{n_0})^{b^n} \end{aligned} \tag{5.40}$$

where $b = i_{\min} - 1$. Applying the upper bound on the BER in (5.34) with $I(X;Y) = I_{n+n_0}$, one obtains

$$\begin{aligned} \text{BER}_{n+n_0} \quad \lesssim \quad & \tfrac{1}{2}\left[(1 - I_{\text{ch.}})\lambda_{i_{\min}}\right]^{\frac{b^n-1}{b-1}} \\ & \cdot \left[\sum_j \rho_j (j-1)\right]^{b\frac{b^n-1}{b-1}} (1 - I_{n_0})^{b^n} . \end{aligned} \tag{5.41}$$

Considering (5.41), one can observe that the behavior of the BER as a function of the number of iterations in the convergence region is *exponential of exponential*. In other words, the BER should exhibit a "threshold behavior," both as a function of the iteration number n and as a function of the minimum variable node degree $i_{\min}$. This result is useful if the goal is to design LDPC codes which achieve very low BER after a small number of iterations, i.e., codes which (i) can be decoded with low complexity, (ii) are characterized by low decoding delay, and (iii) do not need the presence of an algebraic concatenated coding scheme in order to guarantee "near-error-free" performance. This last point is particularly relevant from an application point of view. In fact, if the target BER for an application is, say, 10^{-12} and one designs a code that gives a BER less than or equal to 10^{-12} after 6 iterations and is characterized by a sufficiently large minimum cycle length, the actual BER will be close (or equal) to the predicted one, thus relaxing the need for concatenation with an outer error correction scheme. This also shows that, in order to reach very low BER, a code with low-degree variable nodes needs many more iterations than a code with higher minimum variable node degree. This is the case for codes with performance close to the capacity, which are known to need degree-2 variable nodes [13].

In the following, we perform LDPC code optimization for a small number of iterations and a low target BER. The obtained results shed light on the optimal code structure.

We jointly optimize the degree distributions of variable and check nodes using a slightly modified version of the semi-random walk algorithm presented in Section 5.6. Given a *target* BER and a desired number of iterations, the modified optimization algorithm can be described as follows:

1. the initial degree distributions (properly chosen) are loaded;

2. the channel is worsened (by increasing the erasure probability) until the BER upper bound in (5.34), computed using EXIT charts, is higher than the target BER;

3. new degree distributions (with fixed code rate) are generated considering small (random) perturbation of the previous degree distributions, until the corresponding BER is lower than the previous value;

4. if the maximum number of trials (set a priori) is reached the optimization process stops and the optimized degree distributions are obtained; otherwise, it restarts from step 2.

Considering a number of decoding iterations between 4 and 10 and several values of the target BER, 400 optimization runs were performed for each couple of iteration and target BER values. The obtained variable node degree distributions are obtained by averaging over the optimization results. The following degree distributions were chosen as arbitrary initial set: $\lambda_3 = \lambda_4 = \ldots = \lambda_{12} = 0.1$ and $\rho_6 = \rho_8 = \rho_{10} = \rho_{12} = \rho_{14} = \rho_{16} = \rho_{18} = \rho_{20} = \rho_{22} = \rho_{24} = 0.1$. These distributions correspond to a rate-1/2 code ensemble.

In Figure 5.14, the average optimized variable node degree distribution coefficients are shown as a function of the target BER, considering 5 decoding iterations. One can observe that as the target BER decreases, the fraction of edges connected to high degree variable nodes increases. In other words, a drift towards high degree distributions is observed.

In Figure 5.15, the relationship between the coefficient λ_4, i.e., the fraction of edges in the LDPC code graph connected to degree-4 variable nodes, and the target BER is investigated. Several curves, corresponding to different number of decoding iterations, are shown. Clearly, for any number of iterations, the coefficient λ_4 in the optimized degree distributions reaches a maximum close to 1 and then decays for decreasing values of the target BER—moreover, one can notice that as the number of decoding iterations increases, the maximum moves to the right, suggesting that the evolution towards high degree variable nodes becomes slower.

In order to better understand the structure of an LDPC code designed to converge within a small number of iterations, we investigate the behavior of a function of the variable node degree distribution given by the ratio between

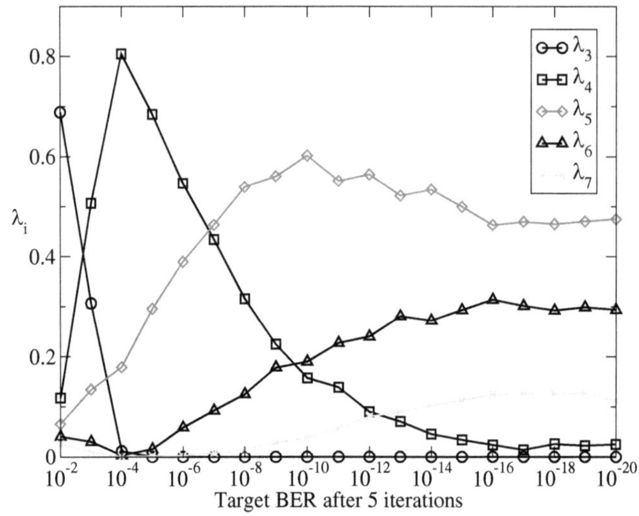

Figure 5.14: Variable node degree distribution coefficients versus target BER considering 5 decoding iterations.

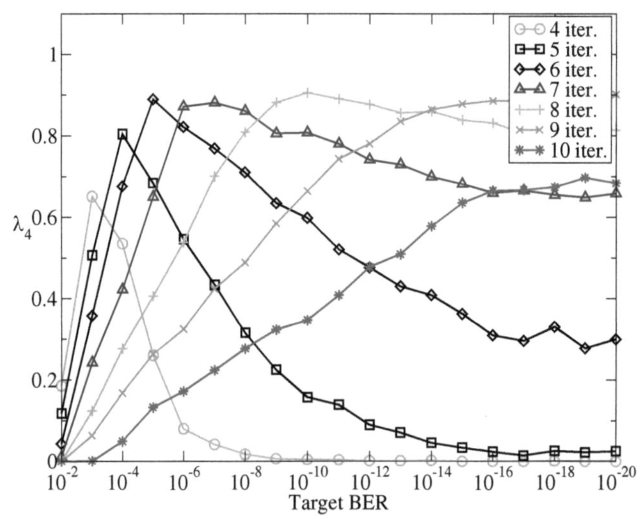

Figure 5.15: Coefficient λ_4 of the variable node degree distribution shown as a function of the target BER. Various decoding iterations are considered.

the number of edges in the graph and the number of variable nodes, i.e., the *average variable node degree* $\bar{d}_v$. This quantity can be computed as follows:

$$
\begin{aligned}
\bar{d}_v &= \frac{\sum_i i n_i}{n} = \frac{\sum_i i \ell \frac{\lambda_i}{i}}{n} \\
&= \frac{\sum_i i \ell \frac{\lambda_i}{i}}{\sum_i \ell \frac{\lambda_i}{i}} = \frac{\sum_i \lambda_i}{\sum_i \frac{\lambda_i}{i}} \\
&= \frac{1}{\sum_i \frac{\lambda_i}{i}}
\end{aligned}
$$

where n_i denotes the fraction of variable nodes of degree i, n is the codeword length, and ℓ is the total number of edges in the graph. The quantity $\bar{d}_v$ is representative of the "connectivity" of the LDPC code bipartite graph where the extremes

$$\bar{d}_v = n(1 - R)$$

and

$$\bar{d}_v = 1$$

represent, respectively, a fully connected bipartite graph and a minimally connected bipartite graph. In Figure 5.16, the relationship of the average variable node degree with the target BER is shown, for various numbers of decoding iterations. The average variable node degree seems to be a monotonically increasing function of the target BER. Moreover, the smaller the number of iterations, the higher the average variable node degree. In Figure 5.17, the relationship between the average variable node degree and the number of iterations is shown at different values of the target BER. The curves are monotonically decreasing. Interestingly, the average variable node degree concentrates in a short range (between 3.3 and 3.7) as the number of iterations increases. More precisely, the trend of the curves suggests asymptotic convergence to a value close to 3.3, *regardless* of the target BER.

Considering 5 decoding iterations and target BERs equal to 10^{-2}, 10^{-6}, 10^{-12}, we have chosen the best obtained degree distributions. The criterion for selecting the best degree distribution is: *among all the codes that satisfy the requirements of BER lower than the target value after 5 iterations, select the one requiring the minimum channel capacity to meet the requirements.*

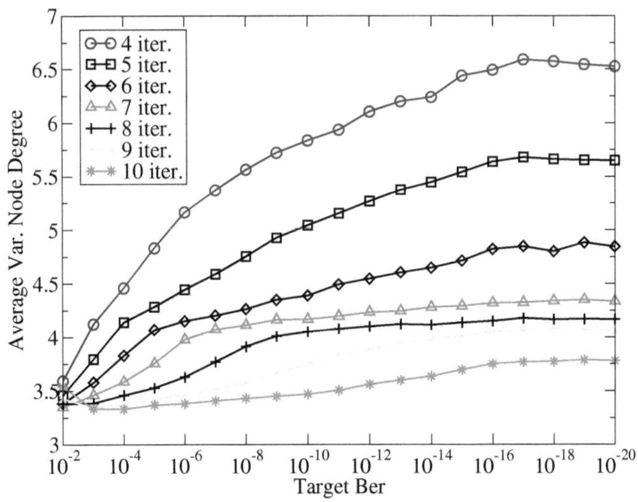

Figure 5.16: LDPC graph connectivity (average variable node degree) versus target BER. Various numbers of decoding iterations are considered.

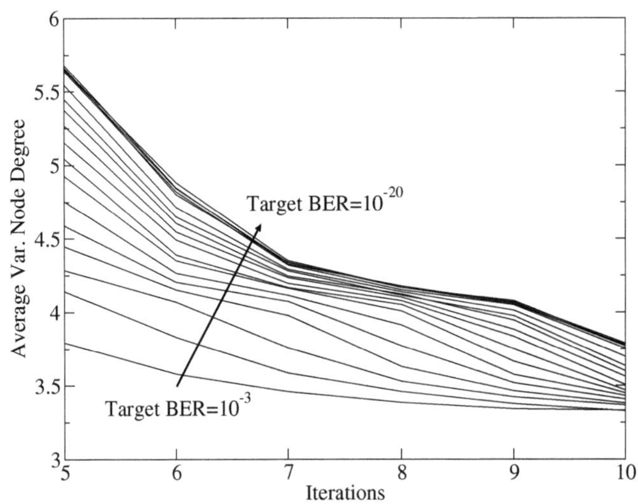

Figure 5.17: LDPC graph connectivity (average variable node degree) versus decoding iteration number. Various values of the target BER $(10^{-3}, 10^{-4}, \ldots, 10^{-20})$ are considered.

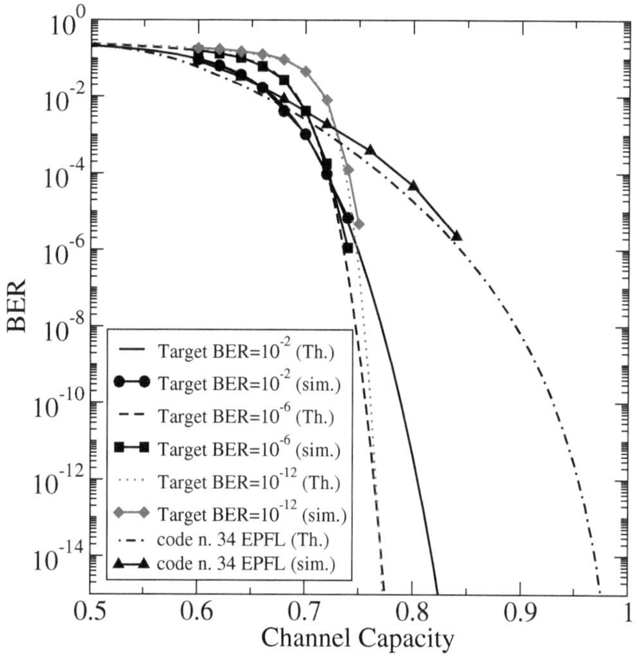

Figure 5.18: Theoretical and simulation-based BER as a function of the BEC channel capacity for 4 codes. The number of decoding iterations is 5.

LDPC codes with codeword length 6000 were then generated considering the selected degree distributions. As a reference, the rate-1/2 degree distributions number 34 in [97] (referred to as "EPFL"), with performance close to the AWGN channel capacity, have been considered, and an actual LDPC code has correspondingly been generated with codeword length 10000. In Figure 5.18, the BER curves of the generated codes are shown, as functions of the BEC capacity $1 - p_e$, where p_e is the probability of erasure. In particular, for each code/degree distribution the figure shows (i) a theoretical BER curve obtained using EXIT charts and the upper bound in (5.34) and (ii) a BER curve obtained by Monte Carlo simulation of the actual code. In all cases, the number of iterations is 5. One can observe that, at BER $= 10^{-2}$ the "ad hoc optimized" code performs better than the codes optimized for target BER equal to 10^{-6} and 10^{-12}. The "EPFL" code exhibits better performance than the code optimized for target BER equal to 10^{-2}; this is due to the fact that in our optimization degree-2 variable nodes were omitted. At a target BER equal to 10^{-6}, the best code is the ad hoc optimized code. One can observe

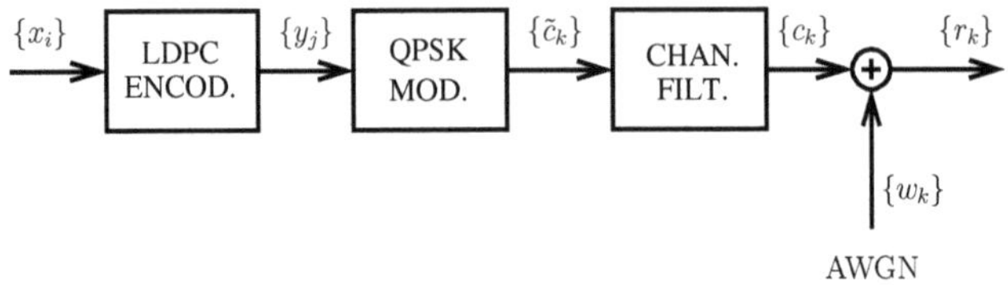

Figure 5.19: Pictorial representation of the transmitter.

that the theoretical BER curve is always lower than or equal to the simulated one, even if the theoretical curve is obtained using the "upper bound." This apparent contradiction is due to the fact that: (i) the actual codes contain cycles and (ii) for the particular case of a BEC the upper bound is achieved.

Interestingly, the variable node degree distributions of the best optimized codes are characterized by the presence of a value $\lambda_{i_{\min}}$ close to 1, i.e., these codes are similar to regular LDPC codes. Moreover, it is clear that codes optimized for a low target BER after a small number of iterations significantly differ from the "high performance" LDPC codes designed to reach the channel capacity. We remark that the presence of short cycles in the code graph is the major concern when a very low target BER is desired. In particular, a low target BER calls for higher degree variable nodes, but the design of an LDPC code without short cycles with high variable node degrees can be troublesome.

5.9 Code Design for Dispersive and Partial Response Channels

5.9.1 System Structure and Analysis

The communication system of interest can be characterized as follows. We refer to the discrete-time equivalent communication model. In Figure 5.19, the transmitter side is shown. An LDPC code encodes bits coming from an information source. The resulting length-N codeword $\{y_i\}$ is subdivided into a sequence of pairs of bits which are encoded into QPSK symbols $\{\tilde{c}_k\}$ using a Gray mapped QPSK modulator. The symbols $\{\tilde{c}_k\}$ are then input to a filter which models pulse shaping and the dispersive behavior of the considered channel. The filtered signal $\{c_k\}$ is then corrupted by AWGN to yield the observables $\{r_k\}$ used by the receiver to perform detection. Since the filter precedes the AWGN insertion, it could be seen as some sort of further encoding performed by the transmitter as in the case of a partial response channel (PRC) transmitter. The receiver is implemented as described in Section 5.3.

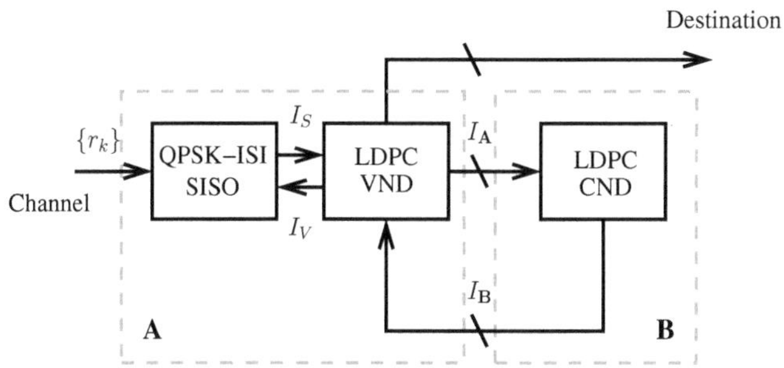

Figure 5.20: Pictorial representation of the receiver.

In Figure 5.20, this receiver scheme is recalled, specializing block **A** with the presence of a SISO block "tailored" for the QPSK/ISI-channel pair. This SISO module implements the FB algorithm for the FSM modeling the considered modulator-channel pair, as described in sections 2.3 and 2.2.2. In Figure 5.21, the FSM structure is shown. The core structure is implemented as a shift register whose input are the pairs of bits forming the QPSK symbol. The length of the shift register is equal to the memory of the ISI channel. At each epoch, the output symbol is obtained by mapping each pair of bits to QPSK symbols, weighing each symbol with the tap weight α_i corresponding to the i-th tap in the ISI channel filter and summing each weighed symbol. As usual, the corresponding SISO module iteratively exchanges soft information with the VND and the CND associated with the LDPC code. Following the notation in Section 5.4, $I_{\mathbf{A}}$ denotes the EXIT curve relative to block **A** (obtained by composing the EXIT curves of the CM-SISO block, denoted as I_S, and the VND, denoted as I_V) and $I_{\mathbf{B}}$ denotes the EXIT curve of block **B**

5.9.2 Theoretical Considerations

In this section, the need for optimized LDPC codes for ISI channels is discussed. In Figure 5.22, various illustrative EXIT curves $I_{\mathbf{A}}$ and $I_{\mathbf{B}}$ are shown. The EXIT curve of block **A** corresponds to a CM-SISO module associated with the concatenation of a memoryless QPSK Gray mapper and an ISI channel. As discussed in Section 5.4, the EXIT curve of block **B** can be assumed not to depend on the particular CM-SISO block, but only on the check node degree distribution of the used LDPC code family.

A possible optimization strategy would consist in modifying these curves, by manipulating the LDPC code degree distributions, in order to guarantee

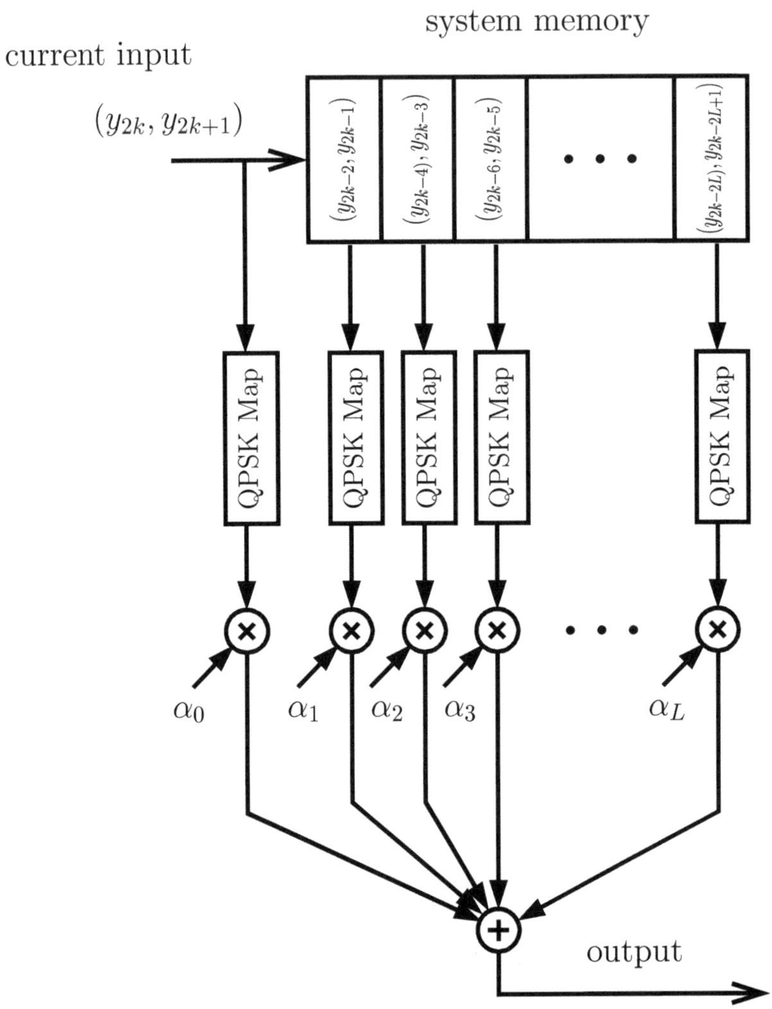

Figure 5.21: Block diagram of the FSM implementation for the QPSK modulator/ISI channel pair.

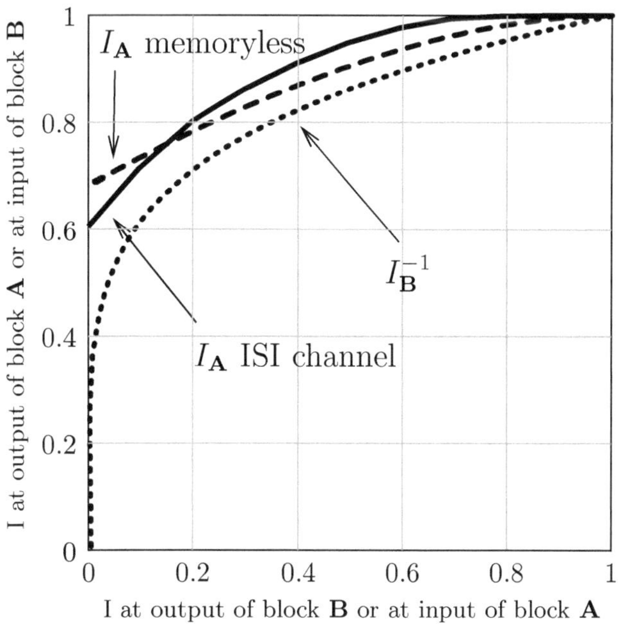

Figure 5.22: Example of EXIT charts for blocks **A** and **B**.

convergence in particular conditions, such as, for example, lowest SNR, minimum number of iterations, etc. This is the case, for example, for the degree distribution optimization technique proposed in Section 5.6.

In Section 5.7, it has been shown that $I_{\mathbf{A}}$ depends on the CM-SISO block EXIT curve $I_S(\cdot)$. In fact $I_{\mathbf{A}} = I_{\mathbf{A}}\big(I_{\mathbf{B}}, I_S(I_V(I_{\mathbf{B}}))\big)$, where $I_V(\cdot)$ and $I_{\mathbf{A}}(\cdot,\cdot)$ depend almost only on the variable node degree distribution. Experience suggests that each one of these functions is monotonically non-decreasing (recall that the underlying SISO algorithms compute *a posteriori* probabilities (APPs) relative to the transmitted bits). Figure 5.12 shows the EXIT curves for several possible CM-SISO modules: a memoryless QPSK demapper for transmission over an AWGN channel and FB algorithms for QPSK transmitted over the ISI channels characterized by the following impulse responses: $(1,1)$, $(1,1,1)$, $(1,2,1)$, $(1,2,3,2,1)$ and $(1,1,0,-1,-1)$. In all cases, Gray mapping is considered and the SNR is set to 2 dB. One can observe that, while the memoryless QPSK Gray demapper is characterized by a flat EXIT curve, the CM-SISO EXIT curves associated with the ISI channels are increasing functions of the input MI. In particular, as can be observed in Figure 5.12, the larger the number of significant taps in the channel impulse response, the steeper the CM-SISO EXIT curve I_S. The composition of the CM-SISO EXIT curve for an ISI channel and the VND EXIT curve, i.e., $I_{\mathbf{A}}$, is then expected to differ from the composition of the CM-SISO EXIT curve for a memoryless demapper and the VND EXIT curve. Thus, the optimal LDPC code for an ISI channel should differ significantly from the optimal LDPC code for memoryless Gray QPSK (which is not different from that of memoryless BPSK). From a practical viewpoint, as we will readily see, the difference between (i) the composition of the CM-SISO EXIT curve for a memoryless demapper and the VND EXIT curve and (ii) the composition of the CM-SISO EXIT curve and the VND EXIT curve is very small in the case of short length impulse responses, such as, for example, a $(1, 1)$ ISI channel. At the opposite, this difference becomes significant for longer impulse responses, such as, for example, a $(1, 2, 3, 2, 1)$ ISI channel.

5.9.3 An ISI Channel SISO Property

By looking at Figure 5.12, one can observe that all the curves "touch" when the input MI is equal to 1. This is not surprising, since the fact that the input MI is equal to 1 corresponds to *complete knowledge of all coded bits but the one for which the APP is being computed.* In other words, we are transmitting, with a linear modulation, a single bit over an ISI channel, which, in case of single bit transmission, reduces to an AWGN channel. Therefore, the FB algorithm-based EXIT curves for QPSK (or BPSK) transmitted over any possible ISI channel converge to the same point $I_S(1) = I_{\text{AWGN}}$, where I_{AWGN} is the capacity of an AWGN channel with BPSK input and with noise sample variance equal to that of the considered ISI channel.

5.9.4 LDPC Codes Designed for ISI channels

Following the approach presented in Section 5.6, we perform optimization of degree distributions for three different LDPC coded concatenated schemes: (i) an LDPC coded QPSK transmitted over a channel with impulse response $(1, 1)$, (ii) an LDPC coded QPSK transmitted over a channel with impulse response $(1, 2, 3, 2, 1)$ and (iii) an LDPC coded QPSK transmitted over a memoryless AWGN channel. The code optimized for an AWGN channel will be used as a reference code, since LDPC codes currently in use are usually optimized for a memoryless (e.g., AWGN) channel. We recall that the optimization consists of a semi-random walk in the parametric space of LDPC code degree distributions, in order to obtain the degree distributions which guarantee the lowest possible convergence threshold (i.e., the lowest SNR for which conver-

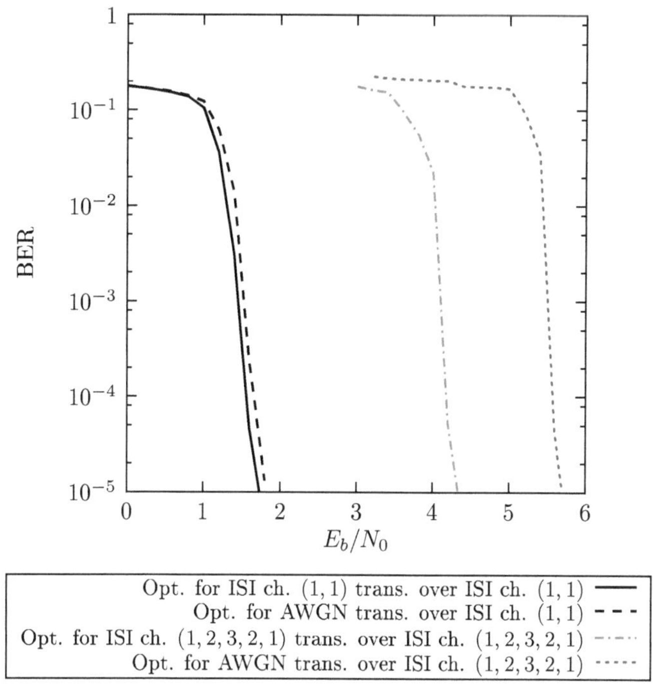

Figure 5.23: BER performance for four systems operating on two different ISI channels: (i) $(1, 1)$ ISI channel and (ii) $(1, 2, 3, 2, 1)$ ISI channel. For each channel both an *ad hoc* optimized LDPC code and an LDPC code optimized for AWGN channel are considered.

gence is guaranteed). The degree distributions are characterized by degrees in the set $\{2, \ldots, 12\}$, in order to allow the construction of moderate length LDPC codes with a small number of short cycles, as discussed in Chapter 3. The optimized codes have rate 0.5. Based on the obtained optimized degree distributions, we construct LDPC codes with codeword length $N = 12000$.

In Figure 5.23, the BER performance of the obtained codes is shown. The LDPC code optimized for AWGN channel is transmitted both over the $(1, 1)$ and $(1, 2, 3, 2, 1)$ ISI channel. For both channels, the performance of the corresponding optimized LDPC code is also shown. Considering the decoder structure parameters presented in Section 5.3, the maximum number of iterations is $N_i = 30$ and $N_{\text{LDPC}} = 30$. Whenever a codeword is found earlier, the decoding process stops. The BER curves exhibit a slight change in convexity, which suggests the appearance of an error floor. This can be attributed to the presence of a small number of short cycles in the LDPC code graph. The presence of this floor is due to the fact that actual codes are extracted at random from given degree distributions for *finite* codeword length—this contradicts the implicit assumption, in the EXIT chart-based analysis, of infinite codeword length.

Considering the BER performance in Figure 5.23, one can observe that the LDPC code optimized for the $(1,1)$ ISI channel leads to a performance very close (i.e., about 0.1 dB) to that of the LDPC code optimized for the AWGN channel and Gray QPSK modulation. As argued in Section 5.9.2, this seems to be due to the fact that the EXIT curves of the CM-SISO block for the $(1,1)$ ISI channel and for the QPSK memoryless channels are similar, thus leading to a very small difference in the EXIT curves of the corresponding block $\mathbf{A}$. On the other hand, considering the BER curves relative to the $(1,2,3,2,1)$ ISI channel, one can readily notice that the performance difference increases to 1.2 dB. This is the result predicted in Section 5.9.2, where we observed that the EXIT curve of the CM-SISO block for the $(1,2,3,2,1)$ ISI channel exhibits higher slope than that relative to the $(1,1)$ ISI channel.

It is possible to compare the codes simulated in Figure 5.23 with other codes proposed in the literature. In particular, in [95, 98] an optimization algorithm is proposed, based on density evolution: the codes are designed in order to exhibit a BER lower than a given constant at a specified number of iterations. Since the underlying analysis model involves infinite dimensional quantities, this optimization algorithm even in an approximate implementation, is computationally very demanding. In [99], a simplified version of the analysis proposed in [95] is given, and used to optimize LDPC code degree distributions for a $1+D$ PRC, i.e., a $(1,1)$ ISI channel: the theoretical threshold of 1.35 dB, obtained in [99] by allowing degree-20 variable nodes, can be compared with the actual codes in Figure 5.23, which exhibit a BER equal to 10^{-3} at an SNR equal to approximately 1.4 dB.

5.9.5 Comments

The results presented in this section can be easily extended to any ISI channel. Nevertheless, by looking at the EXIT curves, relative to the ISI channels with impulse responses $(1,1,1)$ and $(1,2,1)$, shown in Figure 5.12, one could conclude that the steepness of the EXIT curves depends almost only on the number of significant interferers in the impulse response. In particular, channels with short impulse responses exhibit "almost constant" EXIT curves. This finding implies that a standard LDPC code, i.e., an LDPC code designed for a memoryless channel, is a good choice even if the channel were slowly time-varying, as long as the impulse response of the channel is short. On the other hand, it should be noted that ISI channels with long impulse responses exhibit a steeper EXIT curve, and this makes the optimization of LDPC codes worthwhile. These considerations have been verified and confirmed through actual code design and Monte Carlo simulations.

We wish to highlight further simple implications of the results presented in this section. The following facts hold:

- the EXIT curves of the optimal SISO block for a QPSK-ISI channel pair are monotonically increasing;
- they intersect when the input MI (i.e., the *a priori* input to the SISO module) is equal to 1;
- a memoryless channel exhibits a "flat" EXIT curve;
- $I_V(\cdot,\cdot)$ is a monotonically increasing function in both its arguments.

The above facts imply that, with the considered receiver structure, the memoryless channel will exhibit better (or equal) performance than any other ISI channel, regardless of the considered degree distributions. In other words, it is not possible to find a degree distribution pair such that the corresponding LDPC code, transmitted over an ISI channel, entails a performance better than that achieved over an AWGN channel. This will not be the case for DE modulations in Chapter 7—using LDPC codes optimized for DE-PSK over a PSK AWGN channel entails a worse performance than the performance of the same code used over DE-PSK (as expected from and *ad hoc* code optimization).

5.10 Concluding Remarks

In this chapter, we have introduced LDPC coded modulations as a concatenation of an LDPC code with a (possibly coded) modulator suited for the particular communication channel. At the receiver, a SISO block associated with the modulator and the channel is connected to a standard LDPC code decoder, allowing for iterative exchange of information between the LDPC decoder and the SISO for the coded modulation.

The use of EXIT charts for the analysis of the considered receiver schemes has been discussed. This allows to characterize the decoding convergence behavior as a function of the LDPC code degree distributions. A degree distribution optimization algorithm has been illustrated. This algorithm will be used in the next chapters to obtain optimized LDPC codes for memoryless channels and for systems using differential modulation to combat phase uncertainty.

Focusing on the convergence region in an EXIT chart analysis of LDPC coded modulations, i.e., the point $(1, 1)$ of the EXIT chart, the convergence analysis gives useful bounds on the degree distributions, which can be interpreted as a practical generalization to LDPC coded modulations of the bound given in [13] for LDPC codes transmitted over memoryless channels.

We have then used an EXIT chart-based analysis in order to estimate the system BER. This gives the possibility to design LDPC codes satisfying a BER criterion. Codes have been accordingly optimized for a BEC considering a small number of decoding iterations. In particular, insights have been given into the optimized LDPC code structure, showing that LDPC codes designed for low target BER after a small number of iterations (between 3 and 8) are similar to *regular* LDPC codes. The advantage of the proposed approach, with respect to other approaches, such as density evolution-based optimization [13], is that application to other channels, with or without memory, is straightforward.

Finally, as a significant case study, we have considered the design of LDPC coded modulations for transmission over ISI channels or PRCs, illustrating useful insights and practical design guidelines.

Chapter 6

Memoryless Channels and LDPC Codes

6.1 Introduction

As stated in Chapter 1, the introduction of low-density parity-check (LDPC) codes has allowed to achieve near-capacity transmission over some simple channels, such as, for example, the binary input additive white Gaussian noise (BI-AWGN) channel, the binary erasure channel (BEC), or the binary symmetric channel (BSC) [9, 12, 13, 33]. Although the performance of LDPC codes transmitted over binary-input memoryless channels is known and well studied in the literature, a formal proof of their potential to achieve channel capacity is still lacking. Moreover, the application of LDPC codes to generic memoryless channels, and, in particular, for medium to high spectral efficiency signaling over the AWGN channel, represents a promising evolution of established coding techniques, such as, for example, trellis coded modulations (TCM), and have been the subject of attention in the scientific community.

In this chapter, we address the performance of LDPC codes transmitted through a memoryless channel, as described in the following paragraphs.

In Section 6.2, we consider an extrinsic information transfer (EXIT) chart-based analysis of the convergence of the belief propagation decoding algorithm of LDPC codes [69, 76, 77] for binary input memoryless channels. An in-depth investigation of this analysis technique suggests that the performance of LDPC codes depends marginally on the characteristics of the particular memoryless channel, whereas it is dominated by the mutual information (MI) between the input and the output of the channel. Related work also appears in [100–102]. In the following, we will refer to this MI as *constrained input*

M. Franceschini, G. Ferrari and R. Raheli, *LDPC Coded Modulations*,
DOI: 10.1007/978-3-540-69457-1_6, © Springer-Verlag Berlin Heidelberg 2009

channel capacity C_{ci}, in order to distinguish it from the MI used in EXIT chart-based analyses, which refers to the MI between the generic message in the graph and the corresponding codeword bit and is used to track the convergence of the decoding process. Each investigated communication scheme is characterized in terms of the parameter C_{ci}. Moreover, a uniform input binary distribution is assumed, i.e., the *a priori* distribution of an information bit X is such that $P\{X = 1\} = P\{X = 0\} = 1/2$. We will also assume uniform distribution for the LDPC encoded bits, a property that in general holds for every codeword bit in a binary linear code whose generator matrix does not contain all-zero lines.

In Section 6.3, we will use the results obtained in the first part of this chapter to derive an efficient design algorithm for multilevel coding (MLC) [103–105]. The proposed algorithm exploits the partition into memoryless sub-channels induced by MLC, and selects, from a given library of LDPC codes, a subset of codes. The selected subset of LDPC codes is tailored for MLC, and is optimal, in the sense that it maximizes the spectral efficiency guaranteeing a fixed bit error rate (BER) performance above a given SNR. This technique can therefore be used as a practical tool to achieve high spectral efficiency using LDPC codes.

6.2 Performance of LDPC Codes on Binary-Input Memoryless Channels

The assumptions considered in [9] for the derivation of the Gallager A, B, and C iterative decoding algorithms for LDPC codes are valid for a binary input *memoryless* channel. In [69–71], practical approximations of EXIT charts are proposed and used for LDPC code design. Tight upper and lower bounds for EXIT charts have been derived in [62–64, 106, 107]. These bounds allow to find transmission conditions, in terms of MI between the input and the output of the channel, for which it is possible to guarantee convergence regardless of the specific channel. Nevertheless, these bounds do not completely reflect the actual behavior of the decoding process of LDPC codes, which, as the number of iterations increases, seems to converge to that of the BEC bound regardless of the specific channel [93]. This behavior, which has been experimentally observed, seems related to the fact that in the last iterations the BER is low.

In the following, we show that a performance analysis of LDPC codes based on EXIT charts suggests that the behavior of an *ensemble* of LDPC codes (i.e., a set of codes with given degree distributions) does not depend appreciably on the particular memoryless channel, but only on the MI between the input and the output of the channel. This would imply that a code from a given ensemble will exhibit *similar* convergence threshold and performance on any memoryless channel—in other words, different memoryless channels exhibit minor performance differences for a given value of MI. This observation is supported by simulation results relative to various LDPC codes and several memoryless channels. The considered channels are both *symmetric*—binary-input AWGN channel, BSC and BEC—and *asymmetric*—binary asymmetric channel (BAC) and Z channel (ZC). This confirms the early remark in [13], where it was observed that LDPC codes optimized for the AWGN channel show good performance for other memoryless channels, such as BSC and BEC. Similar conclusions where drawn in [100, 101], where the authors show that the performance of LDPC codes over any Gaussian channel, not necessarily AWGN, depends only on the MI between the input and the output of the channel. We remark that the EXIT chart-based analysis of LDPC codes, interpreted as functions of their degree distributions, underlies the implicit assumption that the graphs of the corresponding LDPC codes do not contain short cycles. This condition can be achieved, for example, by choosing a sufficiently long codeword length.

6.2.1 The Start Point in EXIT Charts

In Chapter 4, we introduced a statistical characterization of the convergence behaviour of LDPC coded schemes on the basis of EXIT charts. We now ask ourselves the following question: what is the meaning of the start point of the decoding trajectory in an EXIT chart? The messages at the output of variable nodes at the very first iteration correspond to the logarithmic likelihood ratios (LLRs), based on channel observations, of the transmitted symbols [9]. These quantities are *sufficient statistics* for an optimal decision on the transmitted sequence. This means that the MI between the transmitted binary sequence and these LLRs is equal to the MI between the transmitted binary sequence and the channel output. Since the transmitted bits are assumed to be 0 or 1 with probability 1/2, this MI can also be interpreted as constrained-input channel capacity $\mathrm{C_{ci}}$. Hence, at the first iteration, the MI generated at the output of the variable node detector (VND) is $I_V = \mathrm{C_{ci}}$. As stated in Section 5.4, this value corresponds to the start point of the EXIT chart decoding trajectory in Figure 5.4.

Since simple EXIT chart-based analyses assume that the MI at the output of a block is independent of the particular distribution of the messages, but depends only on the MI at its input, the check node detector (CND) EXIT curve (i.e., the curve I_{B} in Figure 5.4) does not depend on the particular channel.

Nevertheless, the VND EXIT curve may depend on the channel. Experience suggests, however, that the VND EXIT curve (i.e., I_V) depends on the channel through C_{ci} only, whereas it depends weakly on the particular channel type. This is taken into account in the practical approximations currently used for LDPC code design in [69], where the author expresses I_V as a function of the MI between the input and the output of the channel and the MI of the messages coming from the CND. Since, as previously stated, $C_{ci} = I_V(0)$, i.e., the start point of the EXIT chart decoding trajectory, from the considerations above, it follows that the entire function $I_V(I)$ can be characterized by this start point.

Assuming that the EXIT chart-based analysis is accurate, one can conclude that the convergence of the decoding process for an ensemble of LDPC codes, described by their degree distributions, depends only on the constrained-input channel capacity and not on the particular channel. This means that *if* a randomly chosen code of a given ensemble shows, with high probability, a good BER performance when transmitted over a memoryless channel with given C_{ci}, *then* this code, with high probability, will guarantee good BER performance also when used for transmission over any other memoryless channel with equal C_{ci}. The previous consideration is valid *provided that* there is no feedback from the VND to the soft demapper or that the presence of this feedback would not change the MI between the message set at the output of the soft demapper and the generic codeword bit. This occurs in binary input memoryless channels and other significant scenarios which will be addressed in Chapter 7.

It is important to note that these considerations rely on an approximated method, and their accuracy is strictly related to the accuracy of the EXIT chart-based analysis. In the next section, simulation results will be presented that allow to understand to what extent LDPC codes belonging to the same ensemble and transmitted over a memoryless channel show similar performance *regardless* of the specific channel.

6.2.2 Numerical Evidence

We consider Monte Carlo simulation-based performance analysis of three LDPC codes transmitted over five different memoryless channels. The considered codes have rates 1/4, 1/2, and 3/4, and are generated starting from the degree distributions, optimized for the binary-input AWGN channel, found in [97]. The used degree distributions (for variable and check nodes) are given in Table 6.1. The codeword length is set to 10000 binary symbols in all cases. The decoding process stops if a codeword is obtained or a maximum allowed

Table 6.1: Variable and check nodes degree distributions for the considered LDPC codes.

Code Rate	Variable Node Degre Distribution		Check Node Degre Distribution	
	i	λ_i	j	ρ_j
$\frac{1}{4}$	2	0.4161610	4	1
	3	0.2355160		
	5	0.0759725		
	6	0.061250		
	7	0.0013665		
	9	0.0158465		
	10	0.1938870		
$\frac{1}{2}$	2	0.272536	7	0.7
	3	0.237552	8	0.3
	4	0.070380		
	10	0.419532		
$\frac{3}{4}$	2	0.201224	16	1
	3	0.276439	8	0.3
	4	0.033386		
	10	0.488951		

number (equal to 100) of iterations is reached. In both cases, a decision on a binary symbol is made according to the final corresponding LLR value of the symbol, computed as the sum of all the messages sent to its corresponding variable node at the last iteration. The receiver is assumed to know the channel statistics. The soft demapper thus computes the exact APP.

Figure 6.1 shows the considered memoryless channels, which comprise three symmetric channels (binary input AWGN channel, BSC and BEC) and two asymmetric channels (BAC and ZC). We evaluate the BER performance of each considered code over each channel as a function of the constrained-input capacity. For the BAC, the transition probability $P\{0 \to 1\}$ is different from the transition probability $P\{1 \to 0\}$. Two parameters are then necessary to describe this channel (and to compute $\mathrm{C_{ci}}$). We choose to specify the ratio $t \triangleq P\{0 \to 1\}/P\{1 \to 0\}$ as a given constant parameter. The ZC can be interpreted as a particular instance of the BAC with $t = 0$. It is then possible to express $\mathrm{C_{ci}}$ for every channel as a function of a single parameter, namely the SNR γ for the AWGN channel, the transition probability for the

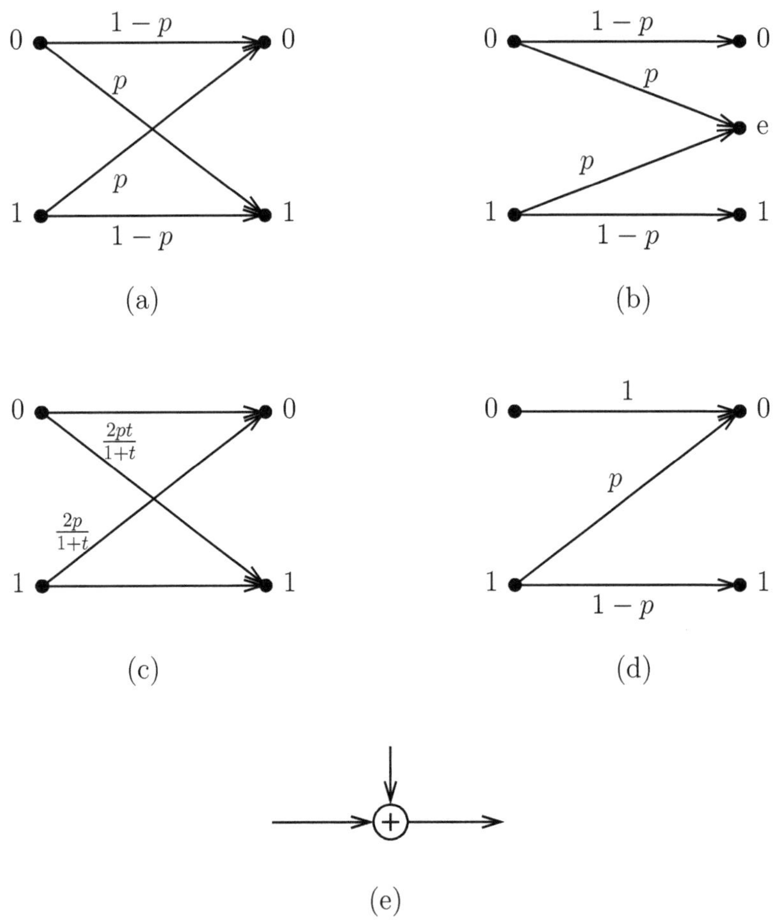

Figure 6.1: Pictorial representation of the considered memoryless channels: (a) the BSC, (b) the BEC, (c) the BAC, (d) the ZC and (e) the BI-AWGN.

BSC, the erasure probability for the BEC, the average transition probability $(P\{1 \to 0\} + P\{0 \to 1\})/2$ for both BAC and ZC (all these probabilities are denoted by p in the following expressions). Summarizing, the constrained-input capacities of the considered channels can be obtained using standard methods and have the following expressions:

$$
\begin{aligned}
\mathrm{C}_{\mathrm{ci}}^{\mathrm{AWGN}} &= \frac{1}{2}\int_{-\infty}^{\infty} \frac{1}{\sqrt{2\pi\gamma}} e^{-\frac{(y-2\gamma)^2}{8\gamma}} \log_2 \frac{2}{1+e^{-y}}\, dy && (6.1)\\
\mathrm{C}_{\mathrm{ci}}^{\mathrm{BSC}} &= 1 + p\log(p) + (1-p)\log(1-p) && (6.2)\\
\mathrm{C}_{\mathrm{ci}}^{\mathrm{BEC}} &= 1 - p && (6.3)\\
\mathrm{C}_{\mathrm{ci}|t}^{\mathrm{BAC}} &= \frac{1}{2(1+t)}\Big\{2(1+t) + pt(1+t)\log\left[\frac{pt(1+t)}{(1+t)(1-p+pt)}\right] \\
&\quad + (1+t)(1-p)\log\left[\frac{(1+t)(1-p)}{(1+t)(1-p+pt)}\right] && (6.4)\\
&\quad + p(1+t)\log\left[\frac{p(1+t)}{(1+t)(1+p-pt)}\right] \\
&\quad + (1-pt)(1+t)\log\left[\frac{(1+t)(1-pt)}{(1+t)(1+p-pt)}\right]\Big\} && (6.5)\\
\mathrm{C}_{\mathrm{ci}}^{\mathrm{ZC}} &= \mathrm{C}_{\mathrm{ci}|t=0}^{\mathrm{BAC}} = \frac{1}{2}[2 + p\log p - (1+p)\log(1+p)]. && (6.6)
\end{aligned}
$$

We remark that for symmetric channels, C_{ci} equals the unconstrained capacity [1].

In Figure 6.2, the BER curves of all three codes, transmitted over the considered five channels, are shown as functions of C_{ci}—note that C_{ci} can assume values between 0 and 1, since the transmitted symbols are binary. For the BAC, the ratio $t = 3$ is chosen as representative. From the results in Figure 6.2, one can conclude that the convergence threshold, in terms of C_{ci}, basically depends only on the code and, in a very limited way, on the channel. Interestingly, this conjecture holds for asymmetric channels as well. The slight differences between the BER curves relative to a specific code cannot be predicted by the EXIT chart-based analysis. In fact, the BER curves depend on the actual code structure, which may contain short cycles [13], and on the statistical distribution of the LLRs at the output of the channel, which are not taken into account by the EXIT charts. Moreover, by considering Figure 6.2 one can quantify the actual difference between the performance of a code transmitted over the considered channels in terms of small fractions of bits per channel use (within a few hundredths).

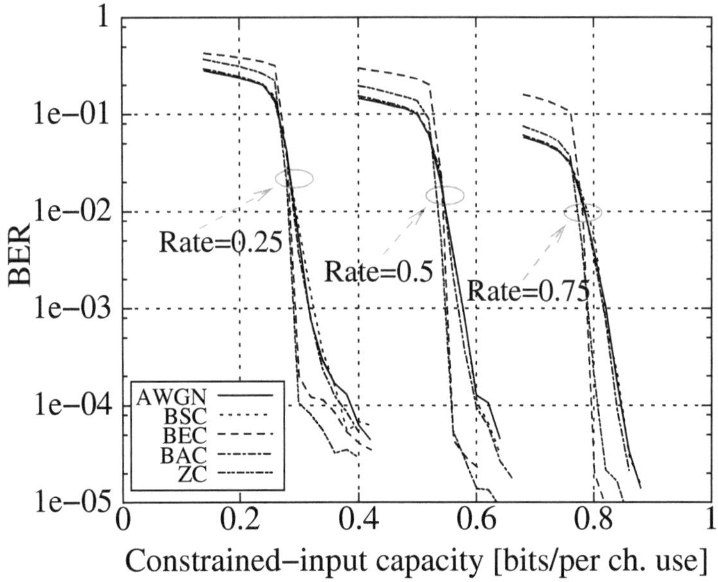

Figure 6.2: BER versus constrained-input capacity for three LDPC codes, characterized by rates 1/4, 1/2 and 3/4, respectively. For each code, the performance for transmission over five memoryless channels is shown. The codeword length is 10000 and the codes are optimized for transmission on the AWGN channel.

6.2.3 Implications

In the previous subsections, we have shown that, with good approximation, LDPC codes which are good for a particular memoryless channel are also good for any memoryless channel, in the sense that they guarantee similar BER performance in the same C_{ci} region, regardless of the channel type.

This allows a characterization of the coding gain in terms of bits per channel use. Since slight differences between different channels have been observed, one can expect that a given LDPC code will exhibit equal performance, within a small fraction of bits per channel use, over different memoryless channels. More rigorous claims regarding our conjecture would involve the derivation of new bounds on EXIT curves of LDPC codes, which extend the results in [62–64, 106, 107], taking into account how the distribution of the messages varies at each iteration. The fact that the performance of an LDPC code has a small dependence on the particular channel has some interesting implications, described in the following.

LDPC Codes Libraries

The independence of the LDPC code performance from the particular memoryless channel implies that good LDPC codes for memoryless channels could be collected in code libraries and reused for several different applications, thus separating the tasks of (i) designing LDPC codes and (ii) fitting them to the considered scenario, which may consist of the concatenation of one or more LDPC codes with a high spectrally efficient modulator .

Soft Demapper without Feedback from VND

The presented conjecture may also impact the design of LDPC codes to be used in a bit interleaved coded modulation (BICM) scheme, which maps binary symbols onto high-order modulation formats [81]. At the receiver side, a soft demapper could generate reliability values for the mapped bits to be passed to the LDPC decoder, which would treat them as channel outputs. In this case, the decoding process would not depend on the particular mapper or channel but only on the MI at the output of the soft demapper. Assuming that iterations between demapper and decoder are not performed or are not useful, LDPC codes designed for a simple memoryless channel (e.g., BSC) will be a good choice also if mapped into high-order modulations. This is due to the fact that, in BICM schemes, the presence of the interleaver after the binary encoder transforms the channel, as seen by the encoder/decoder, into a memoryless channel.[1] However, if iterations are allowed, these claims are no longer valid, since at every iteration the LDPC VND operates on a different input from the soft demapper.

LDPC-MLC Design

Another application of the described property is presented in the following section, where it is shown how to select LDPC codes belonging to a library of good codes for memoryless channels, in order to design an LDPC coded MLC scheme.

6.3 Multilevel Code Design

We now use the concepts developed in the previous sections in order to design MLC schemes [103–105]. In particular, given a set of LDPC codes which exhibit good performance on memoryless channels and featuring several code

[1] This holds exactly only if an ideal random interleaver is used.

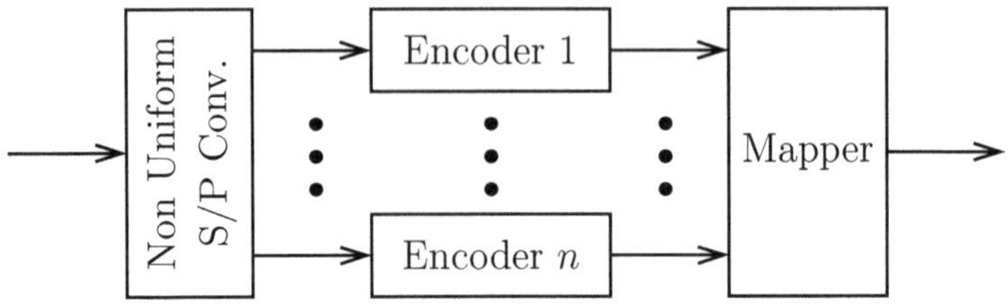

Figure 6.3: MLC transmission scheme.

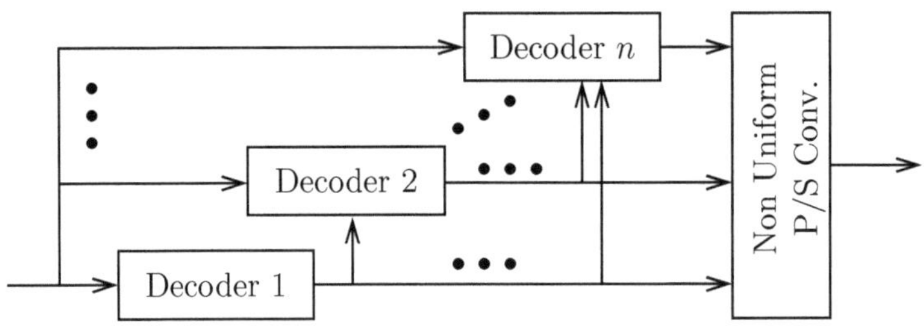

Figure 6.4: Multi-stage decoding scheme: each decoder passes its decisions to every following decoder, which uses this information to compute its own decisions.

rates, i.e., an LDPC code library, we show how to optimally choose the tuple of codes for a MLC scheme operating on a given constellation.

6.3.1 Multilevel Scheme Overview

In Figure 6.3, a MLC transmission scheme is shown for transmission over a 2^n-point constellation. The information bits are split into n different streams by a serial to parallel (S/P) conversion block. Each information bit stream is encoded with a properly chosen LDPC code. The S/P conversion block adjusts the bit rate delivered to each encoder in order to obtain, at the output of each encoder the same binary symbol rate, regardless of the code rate. Hence, the label "Non Uniform S/P." The obtained encoded bit streams feed a mapper which encodes n code bits into a constellation symbol. This coding technique was first proposed in [103], along with a proper decoding algorithm, referred to as multi-stage decoding (MSD). MLC caught growing attention after [105], where it is shown how to determine the rate of each component code in order to retain the full information rate allowed by the considered modulation.

In Figure 6.4, an MSD scheme is shown. For each encoder at the transmitter side, there is a matched decoder at the receiver side. The received observables are fed to each decoder. The decoding process starts at decoder 1, which performs a codeword decision based on the observables from the channels. The decided bits are passed to the remaining decoders. At each stage, the i-th decoder makes a decision on the corresponding codeword, based on the observables from the channel and the codeword decisions corresponding to codes $1, \ldots, i-1$.

Following an information-theoretic viewpoint proposed in [105], we denote the n bits forming a constellation symbol as $X_1, \ldots, X_n$ and the corresponding noisy received observable as Y. The MI between the input and the output of the channel can be expressed as follows:

$$I(X_1, \ldots, X_n; Y) = \sum_{i=1}^{n} I(X_i; Y | X_{i-1}, \ldots, X_1) \tag{6.7}$$

where the right hand side is obtained by applying the chain rule of MI [1]. Equation (6.7) hides a very powerful concept. In fact, it is possible to interpret the stochastic relation between the transmitted bit at the i-th stage X_i and the received observable Y, given exact knowledge of $X_1, \ldots, X_{i-1}$, as a particular memoryless channel. In fact, although it is true that there is dependence between the bits corresponding to a specific constellation symbol, the successive realizations are conditionally independent over time. The overall 2^n-ary input channel can then be seen as as a cascade of channels. Assuming that it is possible to practically achieve "error-free" performance on a memoryless channel through channel coding techniques, one can design a good code for the channel $(X_1; Y)$, then a good code for the channel $(X_2; Y)$ given the knowledge of X_1, and so on. MLC-MSD is the straightforward decoding solution for this scheme. In particular, for each (sub)channel it is possible to achieve error free performance at any rate below the MI of the (sub)channel (uniform input is assumed). With MLC-MSD, it is therefore possible to achieve error free performance at any rate below the MI $I(X_1, \ldots, X_n; Y)$ of the overall channel. We refer the interested reader to [105], where information theoretic aspects of MLC-MSD are covered in details. Among the possible code design techniques for MLC-MSD, selection of the code rate based on the MI of the (sub)channels is referred to as *MI rule* in [105]. According to this rule, efficient coding can be achieved by choosing for the i-th (sub)channel, $i = 1, \ldots, n$, a powerful (and possibly *ad hoc*) binary code with rate slightly lower than $I(X_i; Y | X_{i-1}, \ldots, X_1)$.

6.3.2 Code Selection with the MI Rule

We now show how the results in Section 6.2 affect MLC design based on the MI rule and LDPC coding. Since, over a memoryless channel, the performance of a particular LDPC code without short cycles in the code graph depends almost only on the MI between the input and the output of the channel, as discussed in Section 6.2, we collect a library of LDPC codes, characterized by various rates.

For simplicity, we restrict ourselves to an AWGN channel and linear modulations with given 2^n-ary constellations. The goal is to select an ordered n-uple of LDPC codes in the code library, in order to obtain an overall MLC whose performance can be considered error-free for SNR values below a given threshold. The obtained MLC should have the highest possible rate.

We start with the following consideration: given a constellation, if it is possible to specify a BER threshold $P_b^\star$ and to guarantee that at each level in the MLC scheme the error rate is below this threshold, *then* by specifying a sufficiently low value of $P_b^\star$, it is possible to obtain arbitrary small BER for the overall MLC scheme. This reasonable assumption has an important impact on the code design algorithm described in the following. In fact, consider the property described in Section 6.2, i.e., that the performance of LDPC codes does not depend on the particular channel but only on the channel MI. Every code will be characterized by its own BER versus MI curve. If two LDPC codes have the same rate, for a specified value of $P_b^\star$ one of them will have a better or equal performance, measured as input MI needed to achieve a BER equal to $P_b^\star$, than the other. If both codes fulfill the required design constraints, there is no point in using the code that requires higher MI to achieve the specified BER equal to $P_b^\star$. Thus, for each code rate in the LDPC code library, it is possible to eliminate all codes but the best one. We can thus assume that in the code library the codes will have different rates.

In Figure 6.5, an 8-PSK constellation with natural bit mapping is shown. In Figure 6.6, the MIs of the three sub-channels, as well as with the overall MI, are shown for an 8-PSK constellation transmitted through an AWGN channel. The MI is shown as a function of the SNR E_s/N_0, where E_s is the average 8-PSK symbol energy and N_0 is the one-sided noise power spectral density. The three curves are monotonically non-decreasing. The horizontal dotted line represents the MI needed for a particular code to achieve a BER equal to $P_b^\star$. One can observe that the first subchannel intersects the horizontal line in the rightmost point, the third in in the leftmost and the second in the middle. The meaning of each intersection is that beyond the corresponding SNR, the considered code will guarantee a BER lower than $P_b^\star$ if used for

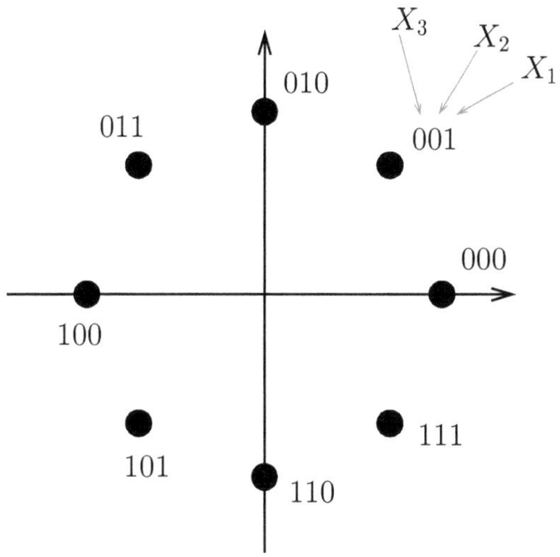

Figure 6.5: Pictorial representation of an 8-PSK constellation with natural mapping.

channel coding on the corresponding sub-channel. If the actual SNR is below that corresponding to the intersection, then the system is in *outage*.

The previous considerations allow one to derive the following MLC design algorithm based on the selection of the proper LDPC codes. The procedure is graphical and the basic graph, shown in Figure 6.7 for a specific scenario (relative to Example 6.1 considered in the following), can be constructed according to the following steps.

- On a graph, plot the MIs of the subchannels versus the SNR. The aggregate MI should be plotted as well.
- For each code in the LDPC code library, plot a horizontal line intersecting the MI axis at a MI equal to the value needed by the LDPC code to obtain a BER equal $P_b^\star$.
- Find the intersection of each code line with each sub-channel MI curve and draw the projection of each intersection on the SNR axis.

In order to find the best code, i.e. the highest-rate code with SNR outage threshold below a given SNR*, it is sufficient, for each sub-channel, to find the intersection of the corresponding MI curve with the code line which has the

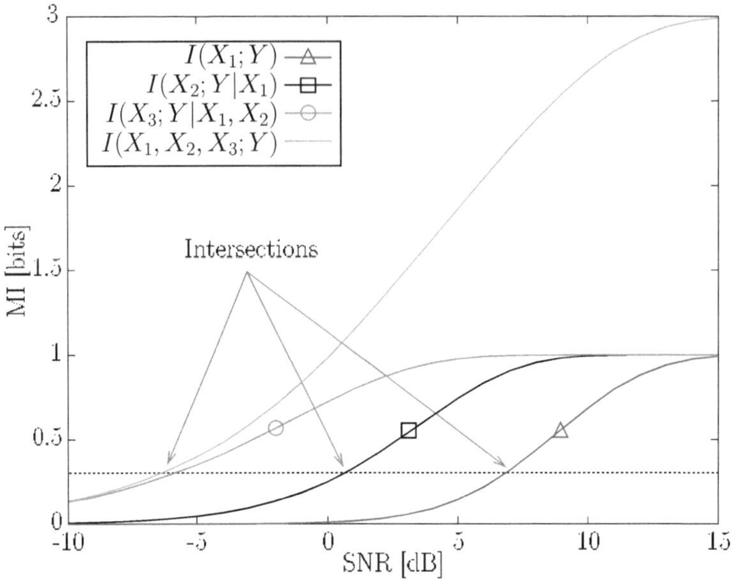

Figure 6.6: MIs $I(X_1;Y)$, $I(X_2;Y|X_1)$ and $I(X_3;Y|X_1,X_2)$ of the subchannels and MI $I(X_1,X_2,X_3;Y)$ of the overall channel for an 8-PSK constellation with natural mapping.

highest SNR below SNR*.

Example 6.1 In Figure 6.7, the proposed graphical algorithm is illustrated for an 8-PSK constellation with natural mapping. Three horizontal lines are plotted, assuming a library which contains only three codes with rates 0.25, 0.57, and 0.88, respectively, for ease of exposition. However, we remark that the approach is general and the use of larger libraries can lead to better results. The intersections are shown on the SNR axis. The intersections corresponding to the first subchannel are marked with triangles, those corresponding to the second subchannel with squares, and those corresponding to the third subchannel are marked with circles. For a given SNR* value, it is sufficent to find the rightmost triangle, the rightmost square and the rightmost circle to the left of this SNR* value. Each symbol, either triangle, circle or square, identifies exactly a subchannel and a code in the code library.

Three LDPC codes were selected using the graph in Figure 6.7 and considering SNR* equal to 10.5 dB, as reported in the figure. The resulting codes can be characterized as follows:

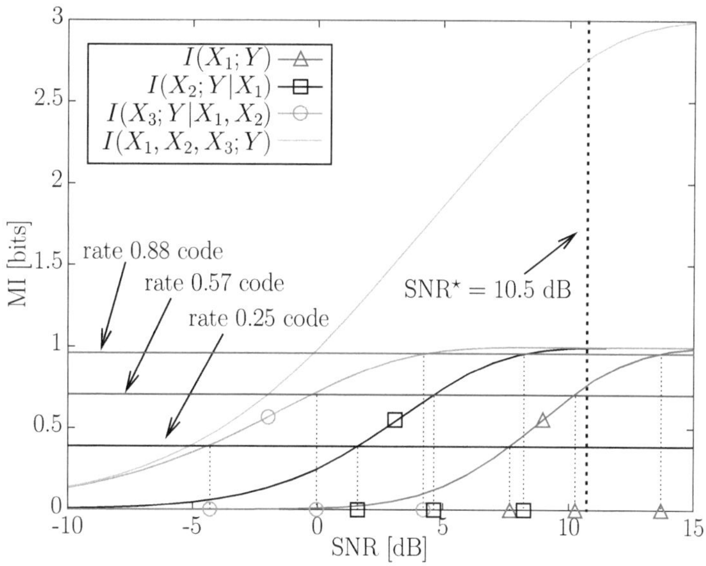

Figure 6.7: Graphical scheme used for MLC code selection.

$$\begin{aligned} \text{channel 1 } (X_1;Y)\text{:} &\quad \text{rate 0.57 code} \\ \text{channel 2 } (X_2;Y|X_1)\text{:} &\quad \text{rate 0.88 code} \\ \text{channel 3 } (X_3;Y|X_1,X_2)\text{:} &\quad \text{rate 0.88 code.} \end{aligned}$$

Had we chosen SNR* equal to 9 dB, the selected codes would have been as follows:

$$\begin{aligned} \text{channel 1 } (X_1;Y)\text{:} &\quad \text{rate 0.25 code} \\ \text{channel 2 } (X_2;Y|X_1)\text{:} &\quad \text{rate 0.88 code} \\ \text{channel 3 } (X_3;Y|X_1,X_2)\text{:} &\quad \text{rate 0.88 code.} \end{aligned}$$

In Figure 6.8, the BER performance of the system designed considering SNR* equal to 10.5 dB is shown. The component codes, with codeword length 10000 and rates 0.57, 0.88, and 0.88, correspond to three regular LDPC codes: $(3,7)$, $(3,25)$, and $(3,25)$, respectively. The overall code rate is $0.57 + 0.88 + 0.88 \simeq 2.33$ bits per channel use. The predicted convergence threshold, i.e., the SNR of the rightmost intersection point within the selected codes, is 10.11 dB, corresponding to a bit SNR $E_b/N_0 = 6.43$. The BER performance of each sub-channel is shown as well. One can observe that there is a good match between the design outage threshold and the actual outage threshold.

We remark that, in order to achieve good performance, the first decoding stages should not introduce errors, since this would affect the next decoding

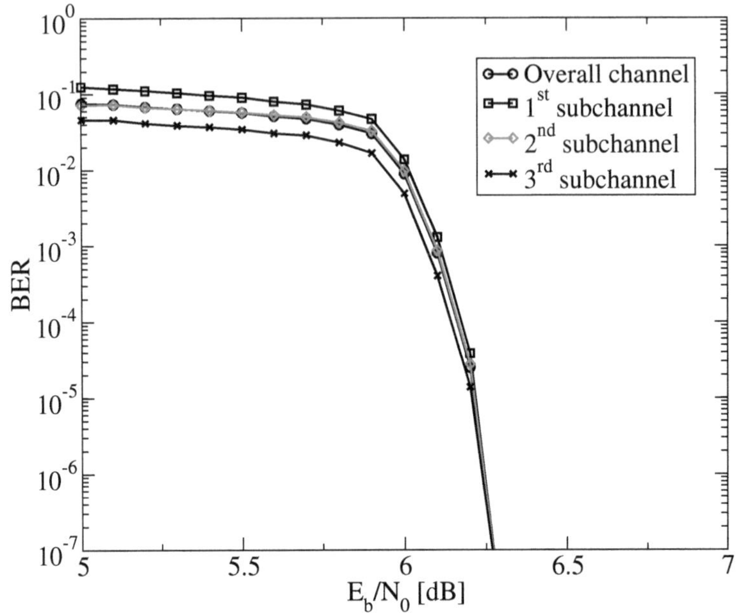

Figure 6.8: BER performance of an 8-PSK MLC-MSD system designed for convergence at SNR equal to 10.11 dB ($E_b/N_0 = 6.43$ dB).

stages, possibly causing avalanche error propagation. The presented numerical results were obtained with *regular* and *quasi*-regular LDPC codes.[2] This choice is due to the fact that the construction of codes without short cycles is easier for regular LDPC codes, especially in the case of low-degree variable and check nodes. The absence of short cycles is useful in order to lower (or make disappear) the error floor characterizing most powerful codes.

6.4 Concluding Remarks

In this chapter, the performance of LDPC codes over memoryless channels has been discussed. The fact that, with a good approximation, the performance of an LDPC code transmitted over a binary-input memoryless channel does not depend on the particular channel but only on the MI between the input and the output of the channel, has been highlighted. This property has several implications, among which that of enabling efficient design of multilevel codes.

[2]We denote as *quasi*-regular LDPC a code with only two, possibly contiguous, allowed variable or check node degrees

In particular, an algorithm for multilevel code design based on the selection of a group of good LDPC codes from an LDPC code library has been described, and some design examples have been given.

Chapter 7

LDPC Codes and Differential Modulations

7.1 Introduction

This chapter investigates the use of differentially encoded (DE) modulations concatenated with low-density parity-check (LDPC) codes. The chapter is logically divided in two parts.

In the first part of this chapter (from Section 7.2 till Section 7.5), we discuss the approach to the design of DE-LDPC coded scheme originally proposed in [79]. Adopting the optimization technique described in Section 5.6, we show a method to design good LDPC codes for DE modulations. We analyze the optimized codes, gaining insights into their graph structures and highlighting the differences between LDPC codes for DE modulations and standard LDPC codes, i.e., optimized for transmission over a *memoryless* channel. We consider the concatenation of an LDPC code with a differential modulator for both phase shift keying (PSK) and quadrature amplitude modulation (QAM). At the receiver side, we make an extrinsic information transfer (EXIT) chart-based system performance evaluation, as described in Section 5.4. We compare the performance of codes optimized for DE modulations with the performance of standard LDPC codes. We show that LDPC codes optimized for DE modulations significantly outperform standard LDPC codes when concatenated with DE modulations. Vice versa, the obtained optimized codes are shown to be tailored *specifically* for the particular DE modulation format and the considered receiver scheme: in other words, while they perform well if used jointly with DE, they perform poorly with memoryless modulation schemes. This will be shown to depend on the presence of a large fraction of degree-2

M. Franceschini, G. Ferrari and R. Raheli, *LDPC Coded Modulations*,
DOI: 10.1007/978-3-540-69457-1_7,

variable nodes.

In the second part of this chapter (Section 7.6 and Section 7.7), we discuss an iterative detection/decoding scheme based on the concatenation of an outer soft-output differential detector and an inner LDPC decoder. The outer detector makes use of a detection strategy, referred to as *detection by multiple trellises* and originally introduced in [108], to perform trellis-based detection over realistic channels. More precisely, we consider channels with unknown parameters and apply the concept of detection by multiple trellises using parallel forward-backward (FB) algorithms (see Chapter 2 for more details). The key idea of our approach consists, *first*, in properly quantizing the channel parameters and, *then*, in considering replicated *coherent* FB algorithms operating on parallel trellises, one per hypothetical quantized value.

7.2 Serial Concatenation of LDPC Codes with PSK and DE-PSK

Consider the transmission side of an LDPC coded modulation scheme described in Chapter 5 and shown in Figure 7.1. As a representative coded modulation (CM) for the transmission system in Figure 7.1, we first consider DE-PSK. For coherent detection, the corresponding CM-SISO module implements, with very low complexity, the FB algorithm. The performance of the considered systems, first studied through an EXIT chart-based analysis, is evaluated in terms of bit error rate (BER) versus bit SNR E_b/N_0, where E_b is the average received energy per bit and N_0 is the one-sided AWGN power spectral density. In all the considered simulations and optimizations, Gray mapping over the PSK constellation is used.

In Figure 7.2 (a), EXIT charts are shown for a regular rate-1/2 $(3,6)$ LDPC code, characterized by the degree distributions $\lambda(x) = x^2$ and $\rho(x) = x^5$. This code without DE, mapped to a quaternary PSK (QPSK) modulation format, is characterized by a good tradeoff, between complexity and performance, for transmission over an AWGN channel. The EXIT curves are computed

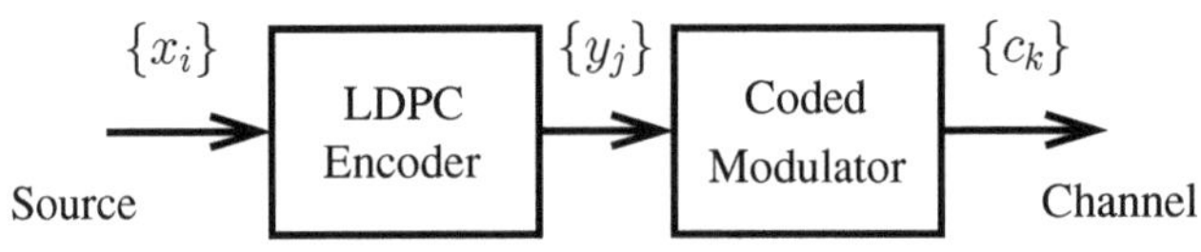

Figure 7.1: System model: transmitter side.

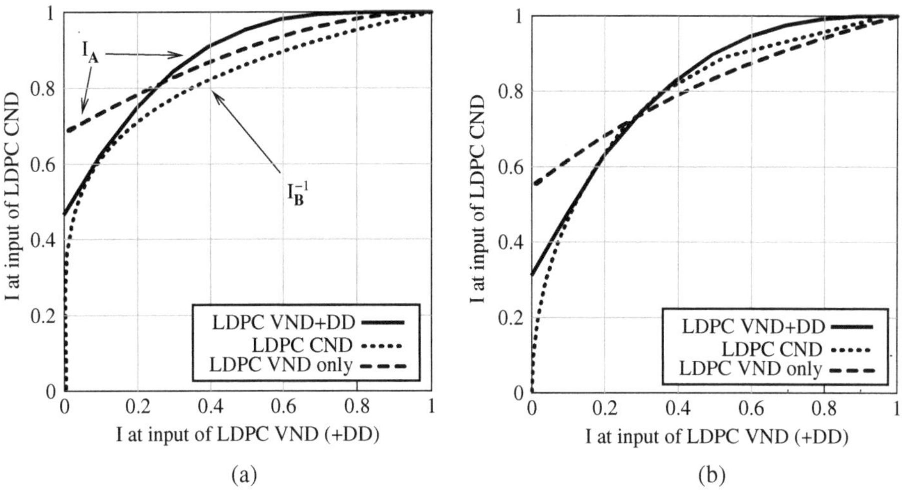

Figure 7.2: EXIT chart-based analysis of a system with serial concatenation of an LDPC code and QPSK: (a) EXIT chart of a $(3, 6)$ regular LDPC code concatenated with a QPSK with DE ($E_b/N_0 = 2.5$dB: tunnel is near *pinch-off*) and QPSK without DE (tunnel is *open*); (b) EXIT chart of an optimized rate 1/2 LDPC code concatenated with a QPSK with DE ($E_b/N_0 = 0.8$dB: tunnel is at *pinch-off*) and QPSK without DE (tunnel is *closed*).

at $E_b/N_0 = 2.5$ dB: the solid curve is the EXIT curve of block **A** (LDPC variable node detector, VND, and differential detector, DD) and the dotted curve is the EXIT curve of block **B** (LDPC check node detector, CND)—for more details on the VND and CND, see Section 5.3. Note that the SNR does not influence the EXIT curve relative to the LDPC CND (the dotted one in Figure 7.2). It is easy to see that the system is at pinch-off: convergence at this and lower values of E_b/N_0 is not possible. The dashed curve represents the EXIT curve of the single LDPC VND: this corresponds to the QPSK system *without* DE, i.e., LDPC BICM. It can be immediately seen that at $E_b/N_0 = 2.5$ dB the tunnel, relative to a transmission scheme without DE, is open. The EXIT chart-based analysis then predicts that, for a bit SNR slightly lower than 2.5 dB, the system with DE does not converge as opposed to the system without DE, which instead converges.

We now apply the optimization technique presented in Section 5.6, forcing the optimization algorithm to use check and variable nodes with specified degree values. As representative values, check nodes of degree 3, 4, 8, and 15, and variable nodes of degree 2, 3 and 4 have been used (these are reasonable choices, but the approach is general). After a few steps, the optimized degree distributions converge to the following:

$$\begin{array}{llll} \rho_3 = 0.3157 & \rho_4 = 0.2259 & \rho_8 = 0.0273 & \rho_{15} = 0.4311 \\ \lambda_2 = 0.5473 & \lambda_3 = 0.0116 & \lambda_4 = 0.4411. & \end{array}$$

Figure 7.2 (b) shows the EXIT curves for this optimized code ensemble for $E_b/N_0 = 0.8$ dB: the solid and dashed curves correspond to block **A** and the dotted curve to block **B**. It is immediate to recognize that the tunnel is at pinch-off. The dashed curve in Figure 7.2 (b) is the EXIT curve of the LDPC VND only (i.e., without DD): the tunnel is "heavily" closed, predicting that the system with DE should perform significantly better than the single LDPC code without DE. Note that the convergence SNR threshold predicted by the results in Figure 7.2 (b) is around 0.9 dB.

In order to closely approximate the degree distributions obtained with the optimization technique, we chose to design LDPC codes with codeword length equal to 6000 binary symbols. In Figure 7.3, the performance of both optimized and regular $(3,6)$ LDPC codes with and without DE is shown. As introduced in Section 5.3, we denote the maximum number of iterations between blocks **A** and **B** as N_i, and to the maximum number of standard LDPC final decoding iterations (between VND and CND) as N_{LDPC}. For DE systems, these maximum numbers of iterations are $N_i = 30$ and $N_{\text{LDPC}} = 30$, whereas for non DE systems a maximum number of 100 standard LDPC iterations is allowed—this makes the complexities of the two different systems very similar. It can be observed that, for a regular $(3,6)$ LDPC code, while good performance is obtained without DE (curve marked with diamonds), the introduction of DE shifts the BER curve to the right, with an SNR loss of about 1.2 dB (curve marked with squares). When the LDPC code is optimized for DE, i.e., block **A** includes a CM-SISO module based on the FB algorithm relative to the DE modulator, it is possible to see the inversion of performance between the system with and without DE, as predicted by the EXIT chart-based analysis. In other words, the use of the LDPC code optimized for DE, in the system with DE (curve marked with triangles) leads to good performance, i.e., it behaves as the regular $(3,6)$ LDPC code without DE (curve with diamonds). On the other hand, the use of the LDPC code optimized for DE, in the system *without* DE, i.e., LDPC BICM (curve with crosses) causes unsatisfactory performance, with an SNR loss of more than 2 dB at a BER equal to 10^{-3}, and low curve slope.

It is also possible to use the CM-SISO module, i.e., the DD, only once and then pass the obtained reliability values to a standard LDPC decoder: the

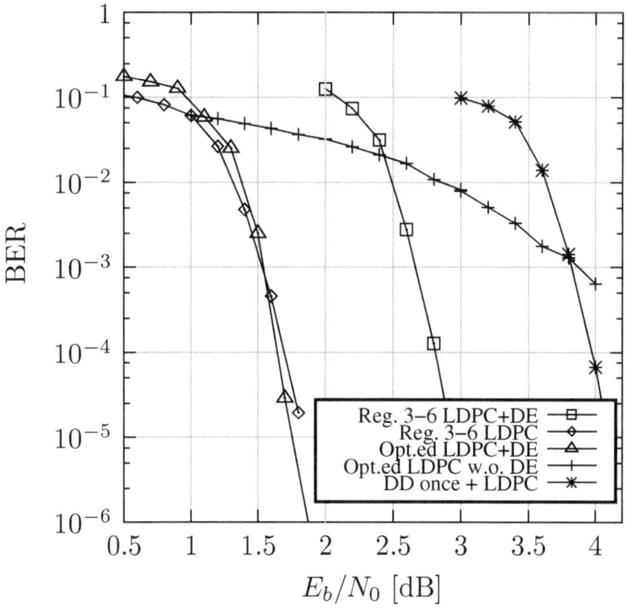

Figure 7.3: BER performance of the four communication schemes considered in Figure 7.2 (a) and (b).

corresponding performance, obtained considering a maximum number of 100 LDPC iterations and using the previous regular (3, 6) LDPC code, is given by the curve marked with stars. It is easy to recognize that the absence of iterations between the CM-SISO block and the LDPC VND leads to a loss of about 1.2 dB with respect to the system with iterative detection/decoding. This can be interpreted noting that the standard LDPC decoder is based on the assumption that a memoryless channel is used, as discussed in Chapters 3–5. When a DE (and the corresponding CM-SISO module) is present, however, the messages passed to the LDPC decoder are significantly correlated. This implies that a large amount of information is embedded in the interdependence of the messages due to the presence of the DE and CM-SISO. The LDPC decoder does not exploit this correlation, thus causing a non-negligible performance degradation. Note that the performance degradation is not due to an ill-conditioned interaction between the correlation structure of the messages and that of the LDPC codewords because the adopted LDPC code is randomly generated, i.e., unstructured.

7.3 Optimized LDPC Codes for PSK

In order to understand the features and limits of the described technique, several optimizations have been carried out, for both a system using DE-PSK and a system using PSK without DE. The set of allowed variable node degrees is $\{2, 3, \ldots, 12\}$, in order to limit the maximum degree and to enable the construction of codes without short cycles. The set of check node degrees is $\{3, 4, 5, 6, 7, 8, 9, 10, 11, 12\}$ for rate-1/2 codes and $\{3, 6, 8, 9, 12, 16, 20, 24\}$ for rate-3/4 codes.

LDPC codes with codeword length equal to 12000 have been designed to match the obtained optimized code ensembles. Each LDPC code has been concatenated with both a PSK modulator and a DE-PSK modulator. In both cases, Monte Carlo simulations have been performed. For the DE scheme, the maximum number of decoding iterations is $N_i = 30$ and $N_{\text{LDPC}} = 30$, whereas for the scheme *without* DE, a maximum number of 100 standard LDPC iterations is allowed. The decoding process stops if a valid codeword is found earlier. In Figure 7.4, the BER curves relative to three LDPC codes optimized for the presence of a DE-PSK modulator are shown: the solid curves are relative to an LDPC code with rate $R = 1/2$ designed for DE-QPSK, the dashed curves are relative to an LDPC code with rate $R = 1/2$ designed for DE-8PSK, and the dotted curves are relative to an LDPC code with rate $R = 3/4$ designed for DE-8PSK. For each LDPC code, the BER curve which exhibits a "cliff" (i.e., the steepest point) at low SNR corresponds to the system for which the LDPC code has been optimized, i.e., the system with DE-PSK (curves marked "with 'DE"); the other curve represents, instead, the performance of the same LDPC code employed in a BICM scheme using PSK modulation with Gray mapping (curves marked "without DE"). For each case, the SNR value corresponding to the capacity limit for the considered coded modulation is shown as a vertical line. The capacity limit for QPSK with code rate 1/2, i.e., with a spectral efficiency of 1 bit per channel use, is 0.17 dB; the capacity limit for 8PSK with code rate 1/2, i.e., with a spectral efficiency equal to 1.5 bit per channel use, is 1.27 dB; the capacity limit for 8PSK with code rate 3/4, i.e., with spectral efficiency equal to 2.25 bit per channel use, is 3.66 dB. All the DE-PSK systems in Figure 7.4 are operating with about $1 \div 1.5$ dB SNR gap to capacity. In other words, the optimized codes guarantee near-capacity performance, even without an exact phase reference.

In Figure 7.5, the performance of LDPC codes optimized for a memoryless PSK modulator is analyzed, both in the presence and absence of DE. For each code, the curve which exhibits a cliff at low SNR corresponds to a system which uses a memoryless PSK modulator (curves marked "without DE"), while

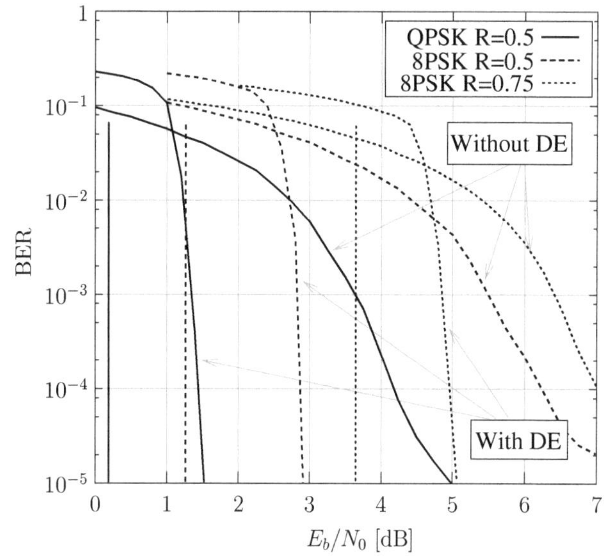

Figure 7.4: Simulated BER for 3 LDPC codes optimized for DE-PSK. Each code is analyzed both with and without DE. In each case, the SNR corresponding to the capacity bound is shown as a vertical line.

the other curve represents the performance of the same code concatenated with a DE-PSK modulator (curves marked "with DE"). The system without DE shows a performance advantage, in terms of SNR corresponding to the cliff of the BER curve, of about 1.5 dB with respect to a system with DE. However, it is important to note that LDPC codes optimized for and used with a memoryless PSK modulator exhibit higher "error floor" with respect to that obtained when the same LDPC codes are used with DE-PSK. The presence of the BER floor in the memoryless PSK modulator is due to the nature of the used code, which contains a small amount of *short* cycles. On the other hand, the absence of the floor in the DE-PSK case can be associated with the fact that the DE-PSK modulator can be interpreted as a rate-1 recursive encoder. As shown in [68], the presence of a rate-1 recursive encoder can reduce short error patterns, responsible for the BER curve flattening, by exploiting the so-called *interleaving gain*.

In Figure 7.6, the coefficients $\{\rho_j\}$ and $\{\lambda_i\}$ of several optimized LDPC codes are shown. Different code ensembles with equal constraints are obtained considering different initial seeds of the pseudo-random number generator embedded in the random walk-based optimization algorithm. The code ensembles

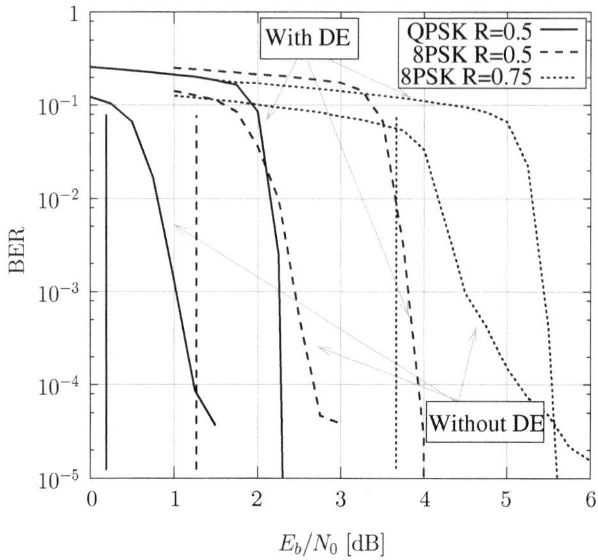

Figure 7.5: Simulated BER for 3 LDPC codes optimized for a memoryless channel. Each code is concatenated with MPSK both with and without DE. In each case, the SNR corresponding to the capacity bound is shown as a vertical line.

in Figure 7.6 (a) and (b) are optimized for DE-QPSK with rate 1/2. The algorithm operates over a limited parametric space, i.e., only a small set of possible node degrees are allowed: the set of variable node degrees is $\{2, 3, \ldots, 12\}$ and the set of check node degrees is $\{3, 4, \ldots, 12\}$. These sets of values makes it possible to design codes without short cycles and reasonable codeword length. The variable node degree distributions $\{\lambda_i\}$ in Figure 7.6 (c) correspond to realizations of rate-1/2 LDPC codes optimized for transmission with BICM PSK. The check node degree distributions appear to give little information, due to the optimization algorithm "residual noise." This is not surprising since, as stated in [13], the performance of LDPC codes exhibit little dependence on the check node degree distribution. Focusing our attention on the coefficients $\{\lambda_i\}$, it is possible to observe that degree-2 variable nodes show a characteristic behavior: in the LDPC code ensembles optimized for DE-QPSK, $\lambda_2 > 0.5$ and $\lambda_2 >> \lambda_i$, $i > 2$. Very similar results, in terms of variable node degree distributions with a predominance of λ_2, were obtained also for LDPC codes optimized for rate-1/2 DE-8PSK and rate-3/4 DE-8PSK. In the LDPC code ensembles optimized for a PSK modulator, λ_2 is still higher than the

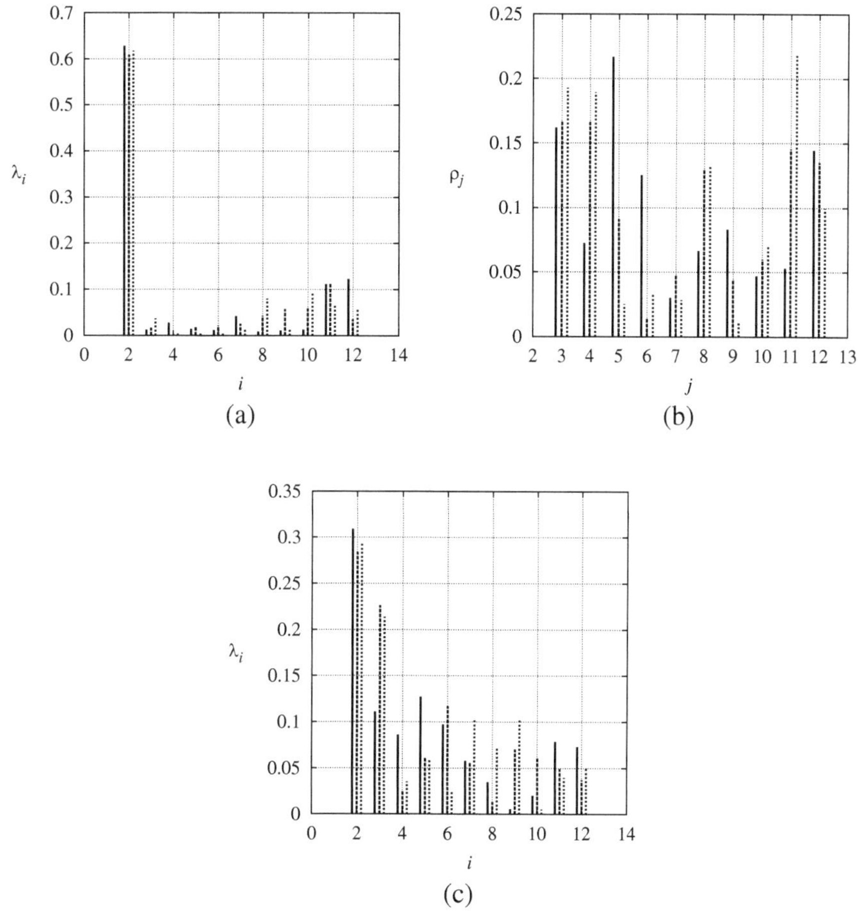

Figure 7.6: Bar diagrams of degree distribution coefficients of three realizations of optimized LDPC code ensembles. In (a) and (b), variable and check node degree distributions for three LDPC codes optimized for rate-1/2 DE-QPSK are shown, respectively. In (c), the variable node degree distributions of three LDPC codes optimized for rate-1/2 QPSK are shown.

other coefficients, but not as high as for DE schemes (see Figure 7.6 (c)).

In [13], a stability condition on λ_2 (inequality (5.30) in Chapter 5) for a standard LDPC decoding algorithm is provided. According to this condition, in order for the BER to approach zero, a necessary condition is $\lambda_2 < \epsilon$, where ϵ is a parameter which depends on the channel and $\rho(x)$. As a reference value, in [13] the authors consider $\epsilon \simeq 0.4$ for a rate-1/2 standard LDPC code. The above discussed results show that this condition is violated if a CM is inserted between the LDPC code and the channel, allowing, in the case of DE-PSK, higher values of λ_2. Note that an LDPC code with a high value of λ_2 is a code whose majority of variable nodes have degree 2, and this corresponds to code graphs with a smaller number of edges (for a given code rate and codeword length). In fact, if ℓ is the total number of edges in the graph and N is the length of the LDPC codeword, it holds

$$N = \ell \sum_i \frac{\lambda_i}{i} .$$

Considering two codes with equal codeword length, variable node degree distributions $\{\lambda_i\}$ and $\{\lambda_i'\}$ and number of edges ℓ and ℓ', respectively, the ratio between the number of edges in the code graph can be written as

$$\frac{\ell'}{\ell} = \frac{\sum_i \frac{\lambda_i}{i}}{\sum_i \frac{\lambda_i'}{i}} . \tag{7.1}$$

If we substitute in (7.1) the degree distributions obtained optimizing for AWGN and for DE, respectively, we obtain a reduction of the edges in the graph of about 20%. Since the computational cost of the decoding algorithm for an LDPC code is proportional to the number of edges in the graph, it follows that LDPC codes optimized for DE-PSK have the pleasant side effect of allowing decoding with lower complexity.

It is generally believed that degree-2 variable nodes exhibit weaker error protection than higher-order variable nodes [13, 34]. However, considering Figure 7.6 (a), one notices that the presence of a large percentage of degree-2 variable nodes is associated with an increase of the fractions of high-degree variable nodes. A possible intuitive interpretation of this behavior is as follows. While a standard LDPC decoder exploits all the available information from the very first iteration, in the considered iterative detector/decoder for LDPC coded modulations the information made available at the "channel input" of the LDPC VND by the CM-SISO block increases with the iterations. This is possible since, at every iteration, the VND passes information to the *a priori*

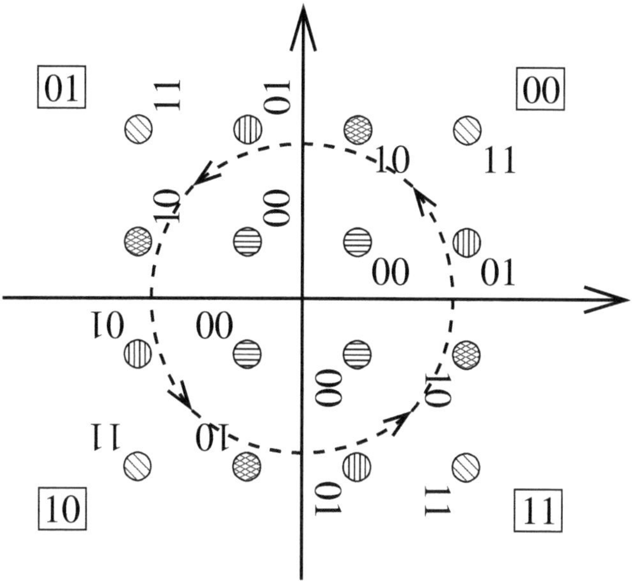

Figure 7.7: Pictorial representation of a DE-16QAM modulation format.

input of the CM-SISO block (see Figure 5.3). Therefore, the critical part of the decoding algorithm corresponds to the first iterations, when the information from CM-SISO block is limited. High-degree variable nodes seem to help the convergence of the iterative algorithm in the first iterations.

7.4 LDPC Codes for DE-QAM

In this subsection, the transmission of LDPC codes concatenated with 16QAM and DE-16QAM is considered. In Figure 7.7, a pictorial representation of a DE-16QAM modulation format is shown. One can observe that two of the four bits at its input are encoded by a Gray mapped DE-QPSK modulator: the obtained point is used to rotate, by an angle equal to a multiple of $\pi/2$, a first-quadrant 16QAM constellation point selected by the other two bits (one bit per dimension).

In Figure 7.8, the BER performance for two communication systems with 16QAM, with and without DE, is shown. Both systems use the same rate-7/8 LDPC code with codeword length 65536. For reference purposes, a vertical dash-dotted line is also shown in correspondence to the capacity SNR, equal to approximately 6.16 dB. The LDPC code is chosen from an ensemble of

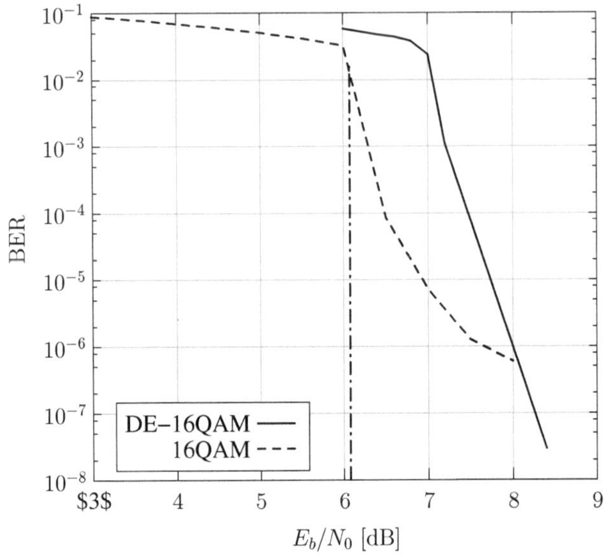

Figure 7.8: BER of a rate-7/8 LDPC code optimized for DE-16QAM and concatenated with DE-16QAM (solid line) and with a 16QAM memoryless modulator (dashed line). The vertical (dash-dotted) line indicates the SNR corresponding to the capacity limit for 16-QAM at the considered code rate.

codes optimized for the presence of DE-16QAM modulator. For DE-QAM, the maximum numbers of iterations are $N_i = 30$ and $N_{\text{LDPC}} = 30$, whereas for QAM without DE a maximum number of 100 standard LDPC iterations is allowed. The solid curve corresponds to a system with a DE-16QAM and the dashed curve corresponds to the system with a Gray mapped 16QAM.

The results in Figure 7.8 show that the code designed for DE-16QAM performs better if used *without* DE. A possible interpretation of this result is that the iteration gain of the DE-QAM SISO module, i.e., the gain enabled by allowing the VND to pass messages to the CM-SISO, is very low. In other words, from an EXIT chart point of view, the considered DE-QAM modulator is similar to a memoryless, Gray-mapped QAM modulator, and this implies that good codes for DE-QAM may also be good codes for QAM. However, the memory introduced by the DE and the corresponding CM-SISO block leads to strong sub-optimality of the processing at the LDPC VND and CND, which assume an underlying memoryless channel. Another immediately noticeable fact is that the BER curve relative to the system without DE is characterized by an error floor, whereas the curve relative to the system with DE does not show any floor in the considered BER range. The floor in the QAM case can be attributed to the presence of a small amount of short cycles in the code graph, which is typical of *random* LDPC codes. Moreover, the differential encoder can be reinterpreted as a rate-1 recursive encoder (at least for the bits which select the quadrant) which, as observed for DE-PSK, is likely to reduce short error patterns.

7.5 LDPC Codes for DE-PSK with Noncoherent Detection

Since the design method described in Section 5.6 can take into account the particular channel, as well as the modulation format and the detection algorithm used in the CM-SISO block, the optimization has been carried out also for LDPC codes concatenated with DE-PSK with noncoherent detection. In the presence of phase uncertainty, the received observation can be modeled as

$$r_k = c_k e^{j\theta} + n_k \tag{7.2}$$

where θ is a random variable constant over the transmitted block and uniformly distributed over $[0, 2\pi)$. While coherent detection can be based on the standard FB algorithm in the CM-SISO module, noncoherent maximum a posteriori (MAP) symbol detection requires some approximations. Following the approach in [109, 110], one can derive a detection algorithm based on a quantization of the phase rotation introduced by the channel. First, the *a posteriori* probability (APP) are computed by the FB algorithm conditionally on one hypothetical channel phase value; then, the conditional APPs are averaged over all possible phase values. The *a posteriori* symbol probability can be written as

$$\begin{aligned} P\{a_k|\boldsymbol{r}\} &\propto P\{a_k\}p(\boldsymbol{r}|a_k) \\ &= P\{a_k\}\int_\theta p(\boldsymbol{r}|a_k,\theta)p_\theta(\theta)\mathrm{d}\theta \end{aligned} \tag{7.3}$$

where $\boldsymbol{r}$ is the vector of all received observations and $\propto$ means that the left member is equal to the right member times a constant independent of a_k. In (7.3), $p(\boldsymbol{r}|a_k,\theta)$ can be interpreted as the extrinsic information generated by a coherent FB algorithm, which assumes a phase rotation θ. The integral in (7.3) can be approximated as a sum over a *properly chosen* discrete set $\mathcal{P}$ of quantized phase values, obtaining:

$$P\{a_k|\boldsymbol{r}\} \mathrel{\tilde{\propto}} P\{a_k\}\sum_{\theta\in\mathcal{P}} p(\boldsymbol{r}|a_k,\theta)P(\theta) \tag{7.4}$$

where $\tilde{\propto}$ denotes an approximation in the relationship described by symbol $\propto$.

Since DE-PSK is insensitive to rotations of the received signal by multiples of $2\pi/M$, where M is the cardinality of the PSK symbols, the set of phases can be a subset of $[0, 2\pi/M)$ [109]. We then choose two possible sets: the first includes 8 equally spaced points (in [110], this is shown to lead to negligible performance degradation), i.e., $\mathcal{P} = \{0, \frac{1}{8}\frac{2\pi}{M}, \ldots, \frac{7}{8}\frac{2\pi}{M}\}$, and the second includes 4 equally spaced points, i.e., $\mathcal{P} = \{0, \frac{1}{4}\frac{2\pi}{M}, \frac{2}{4}\frac{2\pi}{M}, \frac{3}{4}\frac{2\pi}{M}\}$. The optimization algorithm is then run over the same set of node degrees as in the previous subsection. The difference between the degree distributions of LDPC code ensembles optimized for DE-PSK and noncoherent detection and those relative to coherent detection is not noticeable. This is true even if the number of quantization levels used for the computation of (7.4) is reduced to two. An intuitive explanation of this fact is that DE is a technique which makes the communication system insensitive to phase uncertainties, so that the introduction of a further, possibly continuous, phase uncertainty cannot induce a severe system change. Moreover, theoretical results show that, asymptotically, the performance of a noncoherent system approaches that of a coherent system [26, 111–113].

In Figure 7.9, the performance of optimized LDPC codes for DE-PSK with coherent and noncoherent detection is compared. The considered LDPC codes are optimized for DE-QPSK (with rate 1/2) and DE-8PSK (with rates equal to 1/2 and 3/4, respectively); the length of the codeword is 12000 and the maximum allowed numbers of inner and outer iterations are $N_i = 30$ and $N_{\text{LDPC}} = 30$, respectively. The considered numbers of discrete phase values are 8 (curves marked by a triangle) and 4 (curves marked by a square). The curves relative to coherent detection are marked by a circle. It is clear that the phase uncertainty introduces a limited performance loss, as long as the phase quantization is sufficiently fine. Moreover, the results in Figure 7.9 show that, while an 8-level phase quantization introduces negligible performance loss, a 4-level quantization introduces a performance loss of about 0.4 dB. Further analysis on the described noncoherent detection algorithm shows that the number of quantization levels can be reduced to a minimum number of 2, causing a performance loss of about 1.7 dB with respect to coherent detection.

7.6 Detection by Multiple Trellises

In the previous sections, LDPC codes were designed for AWGN and noncoherent channel, i.e., an AWGN channel with a constant and unknown phase uncertainty. Practical channels are often influenced by a number of parame-

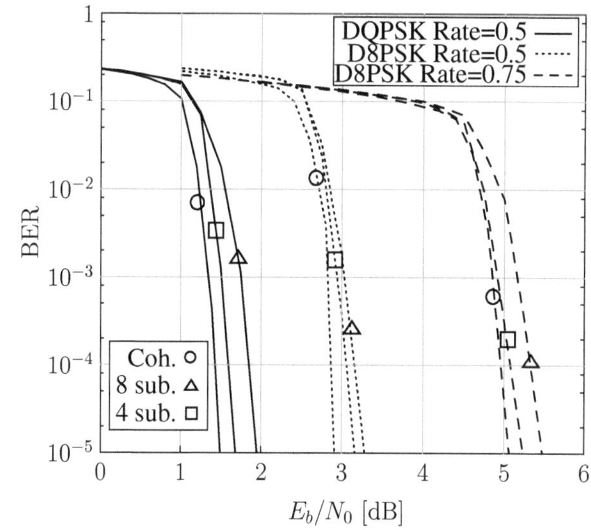

Figure 7.9: BER of LDPC codes optimized for DE-QPSK (rate 1/2) and DE-8PSK (rates 1/2 and 3/4) optimized both for AWGN channel and noncoherent channel.

ters, which in general may vary with time. The problem of designing an effective CM-SISO algorithm in the presence of time-varying parameters is often nontrivial and pursueing optimal solutions might entail a significant computational burden. In this section, a family of CM-SISO algorithms accounting for time varying parameters in the channel/system model are described, with particular emphasis on the phase uncertain channel and the fading channel.

In order to set the problem under study and present the mathematical notation, we begin by reviewing a modified version of the FB algorithm suitable for generic finite-memory channels affected by time-invariant stochastic parameters (see also Chapter 2 for more details on the design of FB algorithms for channels with memory). Afterwards, we will describe the extension to time-varying parameters and analyze two different multi-trellis SISO algorithms.

7.6.1 Time-Invariant Parameters

Let us assume that the channel output is observed for a period of $K + 1$ symbol intervals. The channel can be completely described by the following joint probability density function (pdf)

$$p(\boldsymbol{r}_0^K, \xi | \boldsymbol{a}_0^K) \tag{7.5}$$

where $\boldsymbol{r}_0^K$ (or, simply, $\boldsymbol{r}$) is the vector of the observables $(r_0, \ldots, r_K)$, $\xi \in \mathcal{D}_\xi$ is a stochastic constant channel parameter independent of the transmitted data, $\mathcal{D}_\xi$ is the domain of the channel parameter, and $\boldsymbol{a}_0^K$ is the vector of information symbols a_k transmitted through this channel. Note that (7.5) can take into account possible coding of the information symbol sequence $\{a_k\}$ into a code sequence $\{c_k\}$. We remark that the parameter ξ could be either a scalar parameter or a vector parameter, i.e., ξ could represent a whole set of parameters.

The APP of an information symbol a_k can be expressed as follows:

$$\begin{aligned} P\{a_k | \boldsymbol{r}_0^K\} &\propto p(\boldsymbol{r}_0^K | a_k) P\{a_k\} \\ &= P\{a_k\} \int_{\mathcal{D}_\xi} p(\boldsymbol{r}_0^K | a_k, \xi) . p(\xi) \, \mathrm{d}\xi \end{aligned} \tag{7.6}$$

If, conditionally on the parameter realization ξ, the channel has finite memory [26], the conditional pdf $p(\boldsymbol{r}|a_k, \xi)$ can be computed via a standard FB algorithm [17, 65]. This is possible whenever the transmission system can be modeled as a finite state machine (FSM) whose input and output are, respectively, the information symbol a_k and a random variable (RV) whose statistics depend only on the FSM state and the input symbol (see Chapter 2).

A simple approximation for the computation of the integral in (7.6) is obtained by performing the following finite sum:

$$P\{a_k | \boldsymbol{r}\} \mathrel{\tilde{\propto}} P\{a_k\} \sum_{i=1}^{L} p\left(\boldsymbol{r} | a_k, \xi^{(i)}\right) p(\xi^{(i)}) \tag{7.7}$$

where $\{\xi^{(1)}, \ldots, \xi^{(L)}\}$ is a set of hypothetical quantized values for the channel parameter whose actual values and number L are chosen to obtain the desired accuracy in the numerical integration in (7.6). This corresponds to running L standard FB algorithms in parallel, each one associated with a value $\xi^{(i)}$, $i = 1, \ldots, L$, and computing a weighted average of their outputs to obtain a quantity approximately proportional to the APP.[1]

In the following, we denote the *forward state metrics* computed during the forward recursion of an FB algorithm as $\{\alpha^{(i)}(s_k)\}$, where the superscript

[1] A detailed explanation of the FB algorithm can be found in Chapter 2; further references include [17, 25, 65].

i refers to the FB algorithm associated with the quantized parameter value $\xi^{(i)}$ and s_k denotes the state of the FSM in the corresponding trellis diagram. In particular, we assume that $s_k \in \{0, \ldots, \Xi - 1\}$, where Ξ is the number of states characterizing each trellis. Similarly, we denote the *backward state metrics* associated with the i-th trellis diagram as $\{\beta^{(i)}(s_k)\}$.

Several practical scenarios can be cast within the model described by (7.5), (7.6) and (7.7). In particular, as useful examples, we will consider phase-uncertain and flat fading channels.

Phase-Uncertain Channel

In a communication scenario where the channel introduces a time-invariant phase rotation, the stochastic channel parameter ξ can be equivalently modeled as a phase rotation θ of the transmitted symbol sequence. The discrete-time equivalent observation can be expressed as

$$r_k = c_k e^{j\theta} + n_k \tag{7.8}$$

where r_k is the received observable, c_k is the (possibly encoded) transmitted symbol, and n_k is a (noise) sample of a sequence of independent and identically distributed (i.i.d.) zero mean Gaussian RVs.

Flat Fading Channel

The generic observation model given by (7.5) applies directly to a flat fading channel, provided that ξ has the proper statistical distribution. In particular, in a scenario with unresolvable multipath, ξ corresponds to a fading coefficient f and the channel input-output relation can be expressed as follows:

$$r_k = f\, c_k + n_k \tag{7.9}$$

where, in the case of Rayleigh fading, f has a complex circularly-symmetric Gaussian distribution with zero mean.

7.6.2 Time-Varying Parameters

The idea of detection by multiple trellises stems from an extension of the previous *static-parameter* approach to a scenario with *time-varying* channel parameters.

In order to obtain insights on the impact of the presence of a time-varying parameter, let us consider a useful case study where the channel parameter process $\{\xi_k\}$ is discrete and block constant. Let us assume that ξ_k is uniformly distributed over the set $\{\xi^{(1)}, \ldots, \xi^{(L)}\}$ and constant over blocks of length $N < K$. In other words,

$$\xi_{lN+i} = \xi_{lN+j} \qquad \forall i, j \in \{0, \ldots, N-1\}$$

and the realizations $\{\xi_k\}$ are i.i.d. from block to block, i.e.,

$$p(\xi_{lN}, \xi_{nN}) = p(\xi_{lN})p(\xi_{nN}) = \frac{1}{L^2} \qquad \forall l \neq n\,.$$

As a consequence, the process $\{\xi_k\}$ is a time-varying Markov chain, characterized by an $L \times L$ transition matrix $P_k = (p_{ij}^{(k)})$ at the k-th epoch such that

$$p_{ij}^{(k)} = \begin{cases} \delta_{ij} & \text{if} \quad k \neq N-1 \mod N \\ \dfrac{1}{L} & \text{if} \quad k = N-1 \mod N \end{cases}$$

where δ_{ij} denotes the Kronecker delta. We further assume that the information sequence $\{a_k\}$ is encoded into a code symbol sequence $\{c_k\}$ by means of an FSM. Considering that the observed sequence of length K comprises more than one length-N block with constant channel parameter, the application of a MAP detection strategy to this scenario leads to a time-varying trellis. In Figure 7.10, a representative time-varying trellis for this illustrative block-constant discrete parameter channel is shown. Within a block, i.e., for $N-1$ consecutive time epochs, the trellis structure consists of L "coherent" trellises, each assuming knowledge of ξ, one for each hypothetical quantized value of ξ. In the sections of the various trellis diagrams connecting the states at the end of a block with the states at the beginning of the next block, each state in each coherent trellis is connected with the corresponding state in all the other coherent trellises. In other words, each coherent trellis is connected with any other trellis by the non-zero probability of variation of the parameter value.

A general formulation can be obtained considering an extension of the standard FB algorithm to a channel whose statistics at epoch k are a function of the state ξ_k of a Markov chain. Assume that, given $\{\xi_k\}$, the modulator-channel pair can be described by an FSM, in the sense that the observable statistics are functions of the state σ_k of an FSM whose input is the information symbol sequence $\{a_k\}$. Moreover, let us assume that (i) $\{a_k\}$ and $\{\xi_k\}$ are independent and (ii), given $\{a_k\}$ and $\{\xi_k\}$, the observables are independent. Following the guidelines in [26, 110, 114, 115], it can be shown that the *a posteriori* probability of the symbol a_k can be computed as follows:

$$P\{a_k|\boldsymbol{r}\} = \sum_{(\sigma_k, \sigma_{k+1}):a_k} \beta_{k+1}(\sigma_{k+1})\, \alpha_k(\sigma_k)\, \gamma_k(\sigma_k, \sigma_{k+1}, a_k) \tag{7.10}$$

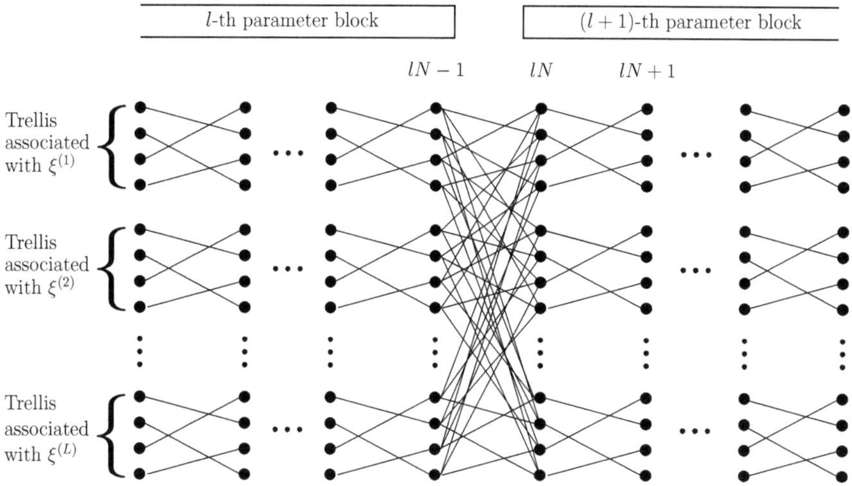

Figure 7.10: Time-varying trellis for detection on block-constant discrete parameter channel.

where, as before, $\boldsymbol{r}$ denotes the vector of the observables and $\sigma_k = (s_k, \xi_k)$ is the (extended) state of the system, the notation $(\sigma_k, \sigma_{k+1}) : a_k$ denotes "the set of all (σ_k, σ_{k+1}) pairs compatible with the input symbol a_k" and the branch "metric" $\gamma_k(\sigma_k, \sigma_{k+1}, a_k)$ is defined as[2]

$$\gamma_k(\sigma_k, \sigma_{k+1}, a_k) = p(r_k|a_k, \xi_k, s_k) \cdot P\{a_k\} \cdot P\{\xi_{k+1}|\xi_k\} \tag{7.11}$$

in which $P\{\xi_{k+1}|\xi_k\}$ is the transition probability between the Markov chain states ξ_k and ξ_{k+1}, and $p(r_k|a_k, \xi_k, s_k)$ is the channel statistical description, i.e., the observable PDF given the data sequence and the channel parameter ξ_k computed at the observation value r_k. The forward and backward "metrics" $\alpha_k(\sigma_k)$ and $\beta_k(\sigma_k)$ are obtained by the following recursions:

$$\begin{aligned}
\alpha_k(\sigma_k) &= \sum_{(\sigma_{k-1}, a_{k-1}):\sigma_k} \alpha_{k-1}(\sigma_{k-1})\, \gamma_{k-1}(\sigma_{k-1}, \sigma_k, a_{k-1}) \\
\beta_k(\sigma_k) &= \sum_{(\sigma_{k+1}, a_k):\sigma_k} \beta_{k+1}(\sigma_{k+1})\, \gamma_k(\sigma_k, \sigma_{k+1}, a_k)\,.
\end{aligned}$$

The FB algorithm in (7.10) operates on a trellis with a number of states equal to the number Ξ of states of the modulator-channel FSM times the number L

[2]Strictly speaking, $\log \gamma_k(\sigma_k, \sigma_{k+1}, a_k)$ is a metric.

of states of the channel parameter Markov chain. This can be interpreted as a "super-trellis" comprising L trellises, each with Ξ states.

As special case, if the Markov chain $\{\xi_k\}$ is time varying and the transition matrix differs from the identity matrix only at time epochs $k = Nl$, with $l \in \mathbb{N}$, it can be easily shown that the forward and backward recursions in the above extended FB algorithm are equivalent to the computation of L independent forward and backward recursions in the Ξ-state trellises for $N-1$ time epochs. Every N time epochs, the recursions involve, in general, all trellises. This corresponds to a block-constant discrete parameter ξ_k, which has been discussed in Section 7.6.2 assuming uniform distribution of the parameter realization. The corresponding super-trellis is pictorially exemplified in Figure 7.10.

Applying the above general formulation, the forward and backward metrics $\alpha_k(s_k, \xi_k)$ and $\beta_k(s_k, \xi_k)$ are functions of the "extended" state $\sigma_k = (s_k, \xi_k)$. They can be computed recursively by running L separate coherent FB algorithms, one for each parameter value. Every N time epochs, in general, $\alpha_k(s_{k+1}, \xi_{k+1})$ and $\beta_k(s_k, \xi_k)$ depend on all forward and backward metrics in all coherent trellises, respectively, i.e., a "mix" of the forward and backward metrics in the coherent FB algorithms is performed. The above considerations can be equivalently drawn by following the guidelines in [110], where a Markov chain model for the channel phase is assumed.

At this point, the idea underlying detection by multiple trellises can be outlined. As for a constant channel parameter ξ, several coherent FB algorithms, characterized by forward and backward metrics $\alpha_k^{(i)}(s_k) = \alpha_k(s_k, \xi^{(i)})$ and $\beta_k^{(i)}(s_k) = \beta_k(s_k, \xi^{(i)})$, respectively, are run independently. The difference with respect to the time-invariant channel parameter case is that every N time epochs, the forward (backward) metrics in the different trellises are properly "mixed" to account for the possible variation of the channel parameter. In the following, we will refer to N as "inter-mix interval."

The idea of considering parallel trellises which occasionally "talk" to each other is appealing, since it is likely to allow both low-complexity and parallel processing. In this sense, performing detection by multiple trellises can be equivalently interpreted as an instance of the *divide et impera* approach to tackle complicated problems with limited complexity.

We remark that the "mixing strategy" should be tailored for the specific communication scenario at hand. Nevertheless, some general considerations can be drawn:

- If ξ is time invariant, the quantity $p(\boldsymbol{r}|a_k, \xi^{(i)})$, computed via a coherent FB algorithm, is expected to be maximum in correspondence to the value

$\xi^{(i)}$ closest to the true[3] channel parameter ξ. In fact, numerical analyses carried out in several scenarios showed that the forward and backward state metrics $\{\alpha_k^{(i)}(s_k)\}$ and $\{\beta_k^{(i)}(s_k)\}$ exhibit an exponential decay in the probability domain. as a function of the epoch k. This is due to the fact that, denoting by $\boldsymbol{\alpha}_k^{(i)}$ the vector of the forward metrics at epoch k in the i-th trellis, the forward recursion can be equivalently expressed as

$$\boldsymbol{\alpha}_k^{(i)} = \Gamma_{k-1}^{(i)} \boldsymbol{\alpha}_{k-1}^{(i)} \tag{7.12}$$

where $\Gamma_k^{(i)}$ is a matrix whose elements are the pdfs of the observable r_k conditioned on every possible transitions in the i-th coherent trellis. In particular, as expected, the decay exponent is higher (i.e., decay is slower) in the FB algorithm associated with the phase value $\xi^{(i)}$ which is closest to the *true* channel parameter ξ, leading to state metrics $\{\alpha_k^{(i)}(s_k)\}$ and $\{\beta_k^{(i)}(s_k)\}$ relatively much larger than those computed by the j-th FB algorithm with $j \neq i$.

- If ξ is time varying, we expect that $\{\alpha_k^{(i)}(s_k)\}$ and $\{\beta_k^{(i)}(s_k)\}$ will try to *adapt* to the parameter changes. This adaptiveness is limited by the fact that state metrics exhibit a "low-pass filter" behavior, i.e., they have *memory* and can change only slowly. This is due to the recursive structure of the metric computation algorithm (7.12). In other words, the FB metric computation process can be equivalently described as a recursive time-varying vector filtering.

- While in standard applications an FB algorithm is insensitive to a possible multiplication of all forward or backward state metrics by a constant, in the algorithm underlying (7.7), the relative weights of different trellises are important. Accordingly, the multi-trellis SISO algorithm turns out to be insensitive to a *normalization* of the metrics *only if* this normalization is carried out, at a given epoch, over all forward or backward state metrics of all parallel FB algorithms.

In the following, two possible "mix" strategies are described.

---

[3] Depending on the symmetry structure of the modulation code, i.e., the law encoding the information symbols a_k into the transmitted symbols c_k, there can be a *set* of ξ values which are optimal, in the sense that they are undistinguishable at the receiver. This may occur, for example, in differential M-PSK transmitted over a phase uncertain channel, where phase rotations of the observed sequences by multiples of $2\pi/M$ cannot be distinguished [109,110].

Multi-Trellis SISO Algorithm 1

At each length-N interval, i.e., at epochs $k = lN$, $l \in \mathbb{N}$, one could manipulate the forward metrics $\{\alpha_k^{(i)}(s_k)\}$ (and, similarly, the backward metrics $\{\beta_k^{(i)}(s_k)\}$) according to the following rule:

$$\alpha_k^{(i)}(s_k) \longleftarrow \sum_{j=1}^{L} \alpha_k^{(j)}(s_k) \quad i = 1, \ldots, L \quad \forall s_k \tag{7.13}$$

where the notation "$\longleftarrow$" represents the assignment of a new value. This corresponds to averaging, for every given state s_k, the metrics relative to all quantized phase values: in other words, the metrics associated with a given state in the various trellises are averaged. We will refer to this algorithm as Algorithm 1. This is the exact APP computation algorithm for the channel with block-constant parameter described at the beginning of Section 7.6.2, if the observables are independent (conditionally on the parameter and the data sequence).

Multi-Trellis SISO Algorithm 2

Assume that the channel is slowly time varying, i.e., assume that ξ_k can exhibit small changes at adjacent epochs. If a suitable manipulation of $\{\alpha_k^{(i)}(s_k)\}$ and $\{\beta_k^{(i)}(s_k)\}$ is allowed only at epochs $k = lN$, with $l \in \mathbb{N}$, the possible transitions of the parameter from one quantization interval to another, occurring amid the block, should be taken into account. Heuristically, it was discovered in [108] that the impact of slow parameter changes within the block can be limited by performing a normalization of the forward state metrics $\{\alpha_k^{(i)}(s_k)\}$ (and, similarly, of the backward state metrics $\{\beta_k^{(i)}(s_k)\}$) as follows:

$$\alpha_k^{(i)}(s_k) \longleftarrow \frac{\alpha_k^{(i)}(s_k)}{\sum_{s'_k} \alpha_k^{(i)}(s'_k)} \quad i = 1, \ldots, L \quad \forall s_k \,. \tag{7.14}$$

where s'_k is a dummy state in the summation, running over all Ξ states of a coherent trellis. This corresponds to a normalization of the state metrics within each FB algorithm, i.e., trellis by trellis, as opposed to a normalization amongst all trellises (as considered in Algorithm 1). We will refer to this algorithm as Algorithm 2.

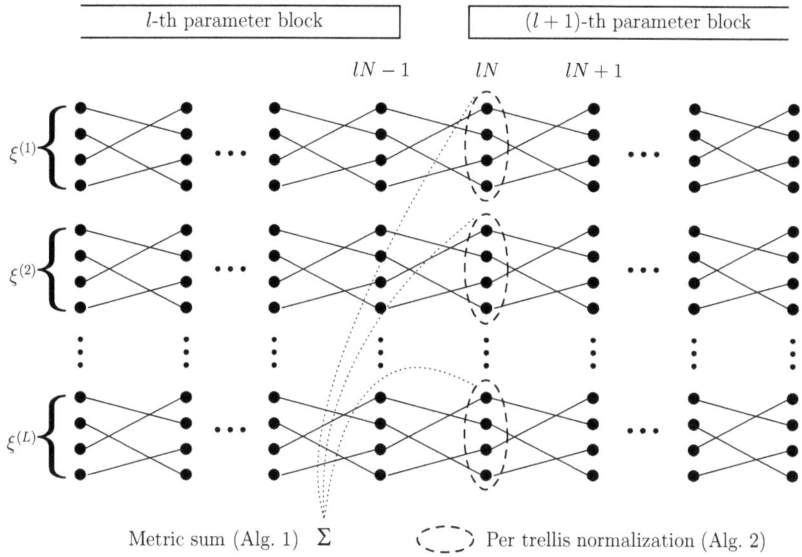

Figure 7.11: Pictorial exemplification of the metric mixes in the two considered algorithms.

Metric Mix in the Algorithms: a Comparison

The manipulations corresponding to (7.13) and (7.14) can be interpreted as a combining or *mixing* of the metrics $\{\alpha_k^{(i)}(s_k)\}$ (similarly for the metrics $\{\beta_k^{(i)}(s_k)\}$). Figure 7.11 gives a pictorial description of the described algorithmic family, highlighting the *metric mix* for both Algorithms 1 and 2. Each depicted trellis diagram is associated with a coherent FB algorithm which assumes a given channel parameter $\xi^{(i)}$, $i = 1, \ldots, L$. The metric mix for Algorithm 1 is shown to "manipulate" the metrics of all trellises summing all metrics on a per-state basis, whereas the metric mix for Algorithm 2 "manipulates" each trellis independently of the other trellises, performing a per-trellis normalization. The mixing epochs $\{lN\}$, i.e., the beginning of the blocks, refer to the forward metric computation. The backward metric computation mix is performed at epochs $\{lN-1\}$.

In both Algorithms 1 and 2, the value of L, i.e., the number of quantized values of the channel parameter, must be chosen considering its impact on both performance and complexity. In particular, by increasing L the performance of the described detection algorithms can be improved, even though for sufficiently large value of L the performance improvement becomes negligible. On the other hand, it can be shown that the complexity of the detection algorithms increases linearly with L [108].

7.7 LDPC Coded Schemes with Detection by Multiple Trellises

7.7.1 Phase-Uncertain Channels

In this subsection, a phase-uncertain channel is considered. First, the algorithms introduced in Section 7.6 are specialized to this type of channel. Then, these algorithms are analyzed and numerical results are given to characterize their performance.

In Section 7.6.1, the model for a channel introducing a time-invariant phase rotation θ is given. In this case, the APP of an information symbol a_k is given by (7.6). Assuming that θ is uniformly distributed, i.e., $p_\theta(\vartheta) = 1/2\pi$ for $\vartheta \in [0, 2\pi)$ (and 0 otherwise), expression (7.7) specializes to the following:

$$P\{a_k|\boldsymbol{r}\} \mathrel{\tilde{\propto}} P\{a_k\} \sum_{i=1}^{L} p(\boldsymbol{r}|a_k, \vartheta^{(i)}) \tag{7.15}$$

where $\{\vartheta^{(1)}, \ldots, \vartheta^{(L)}\}$ is a set of L *properly chosen* phase values [79]. This detection approach for channels with a block-constant random phase was used in [109].

If we assume a slowly varying channel phase (i.e., the bandwidth of the channel parameter process is small compared with the receiver filter bandwidth), the discrete-time observable can be modeled as in (7.8) by incorporating a time-varying phase process $\{\theta_k\}$[4]:

$$r_k = c_k e^{j\theta_k} + n_k \; . \tag{7.16}$$

where $|c_k| = 1$ since DE-QPSK is considered and n_k is a discrete-time complex AWGN process with $\text{Var}\{n_k\} = (RE_\text{b}/N_0)^{-1}$, in which R is the system spectral efficiency in bits per channel use. By suitably modeling the stochastic process $\{\theta_k\}$, one could try to develop an *exact* APP algorithm. Since we do not want to rely on the exact knowledge of the channel parameter statistics, which is seldom available at the receiver, we resort to the multi-trellis SISO algorithms described in Section 7.6.2.

[4]This discrete-time model can be obtained from the continuous-time multiplicative model assuming that the phase process has a bandwidth much smaller than the signal bandwidth.

In this section, we assume that transmission over an AWGN channel is affected by a Wiener phase noise process $\{\theta_k\}$ described by the following recursive relation:

$$\theta_k = \theta_{k-1} + w_k \mod 2\pi \tag{7.17}$$

where $\{w_k\}$ is a sequence of i.i.d zero mean Gaussian variables. The standard deviation of w_k, denoted as σ_θ, is representative of the phase noise intensity.

The adopted LDPC coded modulation scheme uses a regular (3,6) LDPC code. The codeword length is set to 6000 bits. The decoder structure is that described in Chapter 5. The number of inner and outer final iterations is $N_i = 30$ and $N_{\text{LDPC}} = 30$, respectively.

In Figure 7.12, the performance of the described schemes is shown in terms of BER versus SNR. The performance for transmission over an AWGN channel without phase noise, considering an ideal coherent FB algorithm as inner detector, is shown as a reference. The remaining curves show the performance obtained with the considered algorithms. In particular, the curves marked as "Alg1" and "Alg2" correspond to the performance of the schemes with Algorithms 1 and 2, respectively. For each algorithm, several values of the phase noise standard deviation σ_θ (given in degrees in the figure legend) are considered. In each case, the inter-mix interval N is heuristically optimized. The results in Figure 7.12 show that, even in the presence of a significant phase noise (for instance, $\sigma_\theta = 10°$), it is possible to "blindly" process the metrics of the trellises while still achieving an SNR loss as limited as 1 dB. Heuristically, the optimum value of N turns out to be inversely proportional to σ_θ. The results in Figure 7.12 show that Algorithm 2 entails better performance than Algorithm 1. In particular, for very strong phase noise, i.e., $\sigma_\theta = 10°$, Algorithm 1 suffers an SNR penalty larger than 1 dB with respect to Algorithm 2. This is due to the fact that Algorithm 1 completely erases the phase information every N time epochs, whereas Algorithm 2 performs only a "trellis balancing" as described in Section 7.6.2.

In Figure 7.13, a direct comparison between the performance (in terms of BER as a function of the SNR) with Algorithm 1 and Algorithm 2, for a fixed value of the inter-mix distance $N = 15$, and several values of σ_θ, is shown. The value $N = 15$ optimizes the system performance at $\sigma_\theta = 5°$, as shown in Figure 7.12. The remaining system and simulation parameters are those of Figure 7.12. The BER curves show clearly that for values of the phase noise parameter σ_θ lower than or equal to $5°$, decoding convergence is guaranteed for approximately the same value of SNR, whereas if $\sigma_\theta > 5°$ convergence is not guaranteed any longer, i.e., an error floor may appear. In particular, the error floor characterizing the BER curve corresponding to Algorithm 2

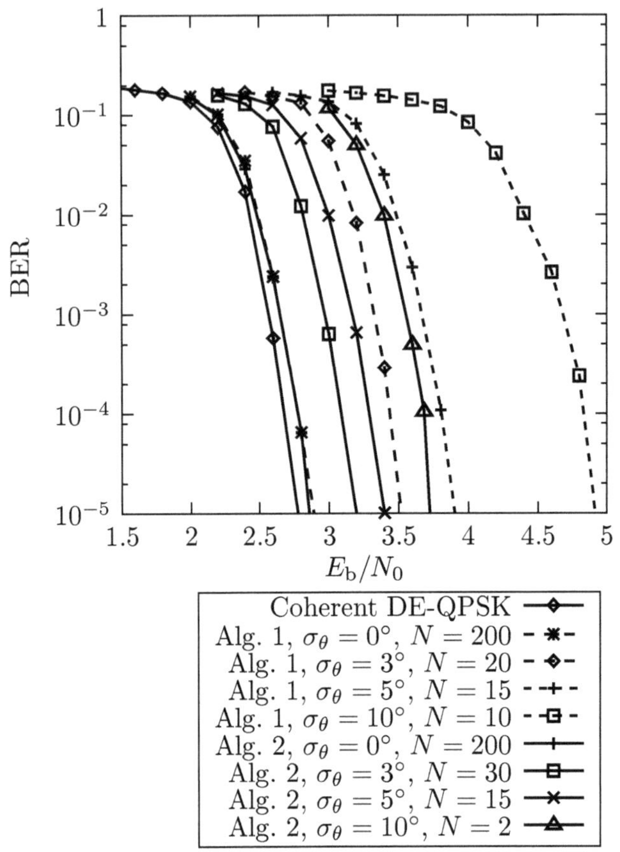

Figure 7.12: BER performance of LDPC coded DE-QPSK schemes based on algorithms 1 and 2.

with $\sigma_\theta = 10°$ is due to the fact that, in order to cope with a strong phase noise, Algorithm 2 needs a very small inter-mix interval N, as clearly shown in Figure 7.12. From the results in Figure 7.13, one can conclude that the described algorithms are *blind* with respect to the phase noise intensity *as long as* this intensity is lower than the value considered in the algorithm design.

7.7.2 Flat Fading Channels

In this section, a flat fading channel is considered. First, we derive the FB algorithm assuming a Markov chain model for the fading channel. Then, we specialize the algorithm introduced in Section 7.6 to the case of flat fading

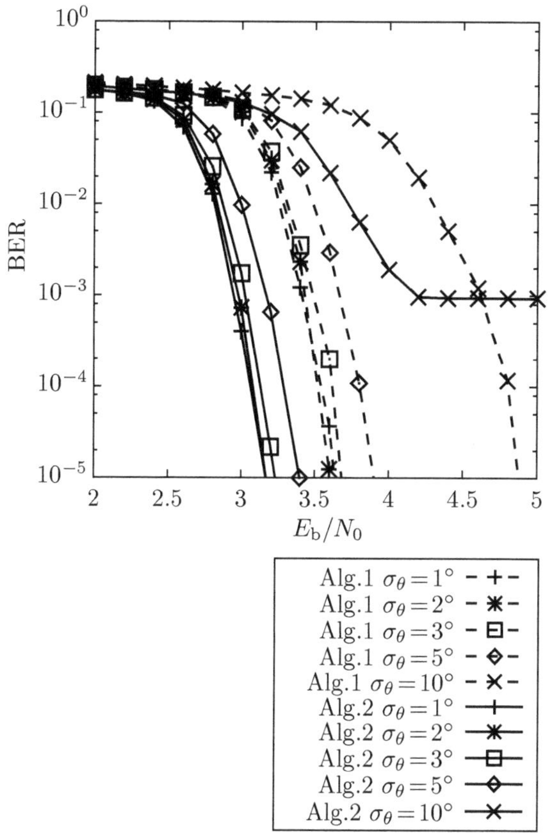

Figure 7.13: BER performance, as a function of the SNR, of the considered Algorithms 1 and 2. Several values of σ_θ are considered and $N = 15$.

channel, highlighting its similarities with the Markov chain-based approach. Finally, the algorithms are analyzed and their performance is characterized through numerical results.

The time-invariant flat fading model given in (7.9) can be extended to a more realistic model with time-varying flat fading. Accordingly, the discrete-time observable can be expressed as

$$r_k = f_k c_k + n_k \tag{7.18}$$

where $\{f_k\}$ is the fading process.[5] In the presence of Rayleigh fading, each realization f_k can be modeled as a zero-mean complex circularly symmetric

[5]We remark that this discrete-time model can be obtained from the continuous-time

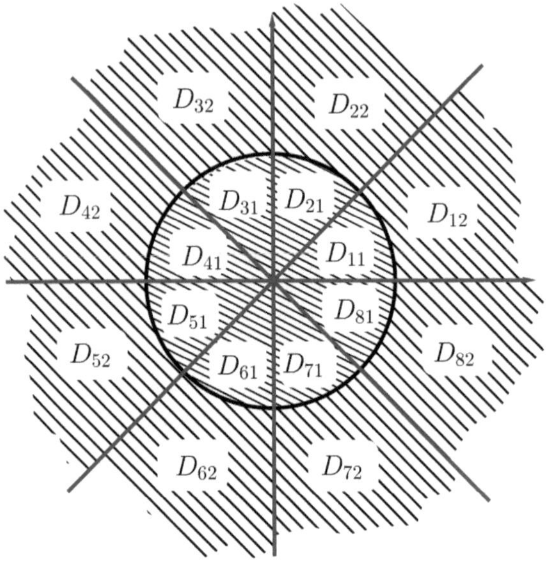

Figure 7.14: Partitioning of the fading complex plane into fading regions.

Gaussian RV. We assume that the fading process $\{f_k\}$ is modeled according to Clarke [116, 117], with zero mean, unit variance and autocorrelation function $R_f(n) = J_0(2\pi n f_D T)$, where $J_0(\cdot)$ is the zero-th order Bessel function and $f_D T$ is the maximum normalized Doppler shift, which characterizes the speed of the fading process.

We now outline the derivation of a simple first-order Markov chain model which approximately describes the evolution of the complex fading process. Several papers deal with Markov chain modeling of the fading process—for more details, we refer the reader to [118, 119] and references therein. We first partition the complex plane into N_{phase} angular sectors $[2\pi\frac{i-1}{N_{\text{phase}}}, 2\pi\frac{i}{N_{\text{phase}}})$, $i = 1, \ldots, N_{\text{phase}}$. Then, we further split each sector into N_{ampl} "ring-shaped" regions. As a consequence, the complex plane is split into $N_{\text{phase}}N_{\text{ampl}}$ subdomains $\{D_{ij}\}$ where D_{ij} denotes the domain corresponding to the i-th phase sector and the j-th ring-shaped region. In Figure 7.14, an illustrative example with $N_{\text{phase}} = 8$ angular sectors and $N_{\text{ampl}} = 2$ ring-shaped regions is shown.

By associating the fading regions with states, it is possible to describe the evolution of the fading process through the use of a Markov chain. In

multiplicative fading model assuming that the fading process has a bandwidth much smaller than the signal bandwidth.

general, considering a first-order Markov modeling for the fading process,[6] the total number of fading states is $L = N_{\text{ampl}} N_{\text{phase}}$. The probabilities of transition through different fading states can be computed by proper numerical integrations. For example, in order to evaluate the probability of transition from the region D_{ij} to the region D_{kl}, one can follow the method in [118], which is accurate as long as the first-order Markov chain modeling of the fading process holds and, in turns, corresponds to a scenario where the fading process is sufficiently *slow* [119].

Since the fading process is modeled through a Markov chain whose state corresponds to the current fading subregion D_{ij}, it is possible to derive a proper FB algorithm for the computation of the APPs of the transmitted symbols $\{a_k\}$. A general formulation accounting for a finite-memory channel depending on a generic process $\{\xi_k\}$ modeled by a Markov chain can be found in [108].

In the following, we will assume that the symbols $\{a_k\}$ are quaternary and encoded by a DE-QPSK encoder before transmission. The channel parameter ξ_k corresponds to the fading region $\hat{f}_k \in \{D_{ij}\} \quad i = 1, \ldots, N_{\text{phase}}$, $j = 1, \ldots, N_{\text{ampl}}$. The extended state described in Section 7.6.2 here is $\sigma_k = (s_k, \tilde{f}_k)$, where s_k is the DE-QPSK encoder state at epoch k, and the fading region $\tilde{f}_k = D_{ij}$ for some i, j, has been substituted to the generic parameter ξ_k.

The two essential ingredients needed for actual implementation of the Markov chain-based SISO algorithm in a scenario with fading are the transition probability $P\{\tilde{f}_{k+1}|\tilde{f}_k\}$ between the Markov chain states $\tilde{f}_k$ and $\tilde{f}_{k+1}$, obtained by suitably modeling the fading Markov chain, and the conditional PDF of the observable $p(r_k|a_k, \tilde{f}_k, s_k)$, given by the following expressions:

$$
\begin{aligned}
p(r_k|a_k, \tilde{f}_k, s_k) &= \frac{p(r_k, \tilde{f}_k|a_k, s_k)}{p\{\tilde{f}_k\}} \\
&= \frac{\displaystyle\int_{\tilde{f}_k} p(r_k|f, a_k, s_k) p_f(f) \mathrm{d}f}{\displaystyle\int_{\tilde{f}_k} p_f(f) \mathrm{d}f}
\end{aligned}
\tag{7.19}
$$

where the independence between the fading process and the DE-QPSK coded data sequence c_k is exploited, $p(r_k|f, a_k, \sigma_k)$ is a Gaussian PDF (with mean fc_k), and $p_f(f)$ is the PDF of the fading coefficient.

[6] We remark that the considered approach can easily be extended to higher-order Markov models of the fading process, at the expense of an increased number of fading states.

The concept of detection by multiple trellises can be now directly applied to a fading channel. In particular, as for the phase-uncertain channel, if the channel is characterized by block-constant fading, Algorithm 1 is an optimum solution. In order to simplify the metric computation, the integral in (7.19) will be approximated by a finite sum of simple Gaussian metrics. However, it was observed that this can lead to numerical problems at high SNR, where the noise variance becomes small. To overcome this problem, one may increase the accuracy of the numerical integration techniques used to compute (7.19) or prevent the variance of the Gaussian pdfs to become too small and trigger numerical problems.

Observe that every concatenated scheme with a powerful error correction code is characterized by a bad BER performance below a given SNR threshold and an operational BER performance beyond this threshold.[7] If the detection algorithm assumes a given, fixed, SNR value, one is guaranteed to obtain optimal performance only when the actual SNR value equals the assumed value. The BER of the fixed-SNR receiver as a function of the SNR, is still expected to be monotonically decreasing. Therefore, if the assumed SNR is fixed to guarantee an operational BER at that very SNR value, the fixed-SNR algorithm will guarantee operational BER beyond this SNR as well. As a consequence, we chose to fix the variance of the Gaussian metric, i.e., the SNR assumed by the detection algorithm, and to make it independent of the actual noise variance. This allows to overcome numerical problems and leads to a completely blind detection algorithm, which does not need either knowledge of fading or noise statistics.

Unlike commonly considered in the literature, where the fading process used in the simulations is generated according to the considered Markov chain model, in the following the fading process used in the simulations is generated according to a realistic Clarke model.

In order to verify the effectiveness of the described detection by multiple trellises, we consider applications to DE-QPSK, both uncoded and coded by a regular (3,6) LDPC code with codeword length 32000—this length allows to counteract long fades. The code should, in fact, "observe" a received sequence long enough to accurately describe the statistics of the channel, i.e., to exploit its ergodicity. We performed simulations considering $N_{\text{ampl}} = 2$ and $N_{\text{phase}} = 16$ and considering Algorithm 1 and the above described simplified metric scheme. Algorithm 2, in the case of fading channel, exhibits

[7] In actual systems, the transition from bad BER performance to operational BER is not perfectly sharp, i.e., it happens within a small SNR region, usually referred to as *waterfall region*.

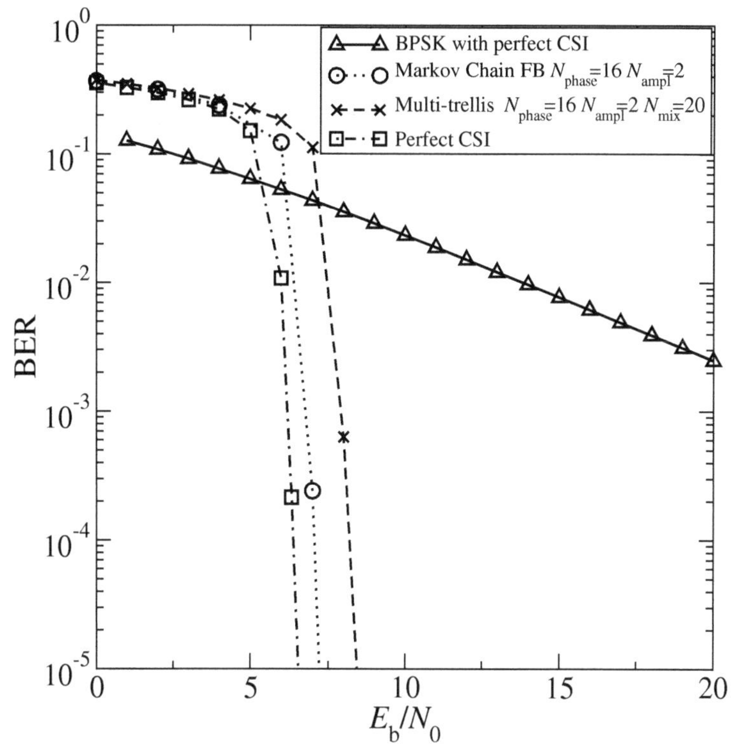

Figure 7.15: BER performance, as a function of the SNR, in a scenario with a flat Rayleigh fading channel. Various schemes are considered: (i) BPSK with perfect CSI, (ii) LDPC coded QPSK with Markov chain-based FB detection, (iii) LDPC coded QPSK with multi-trellis SISO detection, and (iv) LDPC coded QPSK with perfect CSI.

unacceptable performance, and, therefore, is not shown. This is due to the fact that the mix operation in Algorithm 2 normalizes independently every trellis thus assigning large weights to trellises characterized by incorrect fading *amplitudes*. The considered normalized Doppler rate $f_D T$ is equal to 0.01, corresponding to a moderately fast fading channel. The obtained results are shown in Figure 7.15. The multi-trellis curve is obtained assuming a noise variance value corresponding to an SNR of about 7 dB. The inter-mix interval is heuristically optimized by trial and error and set to 20. In every LDPC coded scheme, a number of inner iterations and final LDPC decoder iterations equal to $N_i = 30$ and $N_{\rm LDPC} = 30$, respectively, is used. The Markov chain-based algorithm presented at the beginning of this subsection is also investigated and its performance is shown. As a reference, the performance of (i) the described concatenated scheme and (ii) an uncoded BPSK signaling, both considering perfect CSI, is also shown. As one can immediately see, the performance loss incurred by the use of the described detection by multiple trellises can be quantified at about 1 dB in comparison with the Markov chain model performance and 1.8 dB compared with the perfect CSI scenario.

7.8 Concluding Remarks

In this chapter, LDPC coded modulation for differential modulation schemes has been considered. The two main aspects of LDPC coded modulation design are the design of the LDPC code and the selection/design of the proper modulation and corresponding CM-SISO. The first important consideration is that using standard LDPC codes for memoryless channels in a DE scheme leads to a significant performance loss, which can be avoided using properly designed LDPC codes. A second consideration arises from the analysis of the performance of LDPC coded modulation schemes with multiple trellises detection. Multiple trellis detection is a suboptimal detection which cannot be used in the absence of coding since its use in an uncoded scheme leads to a significant error floor. Nevertheless, in an LDPC coded modulation scheme this remarkable sub-optimality becomes negligible. This highlights the fact that a suboptimal CM-SISO scheme cannot be characterized by its performance in the absence of coding: the use of a concatenated LDPC coded modulation scheme allows powerful simplifications which might be catastrophic in an uncoded scenario.

Chapter 8

Final Remarks

In this book, we have described a method for using low-density parity-check (LDPC) codes for constructing coded modulation schemes for generic communication channels. The basic idea is to use, at the transmitter side, an LDPC code concatenated with a modulator suitable for the particular channel. A good practical choice is to use a modulator whose behavior on the particular channel is well understood and whose practice of use is consolidated. At the receiver side, a soft demodulator, associated with the modulator and accounting for the communication channel statistical behavior, and a standard LDPC decoder iteratively exchange messages. We have shown how to design LDPC codes optimized for the particular transmission scheme. Depending on the specific choice of channel and modulation scheme, the optimized codes might entail remarkable performance gains with respect to standard LDPC codes, i.e., optimized for memoryless channels. It is interesting to note that LDPC codes optimized for a specific scenario are not, in general, good when applied to a different context. For example, a code optimized for the presence of a differential phase shift keying (PSK) modulator and a noncoherent channel has extremely poor performance if used jointly with a (non-differential) binary PSK (BPSK) modulator for transmission over an additive white Gaussian noise (AWGN) channel.

With the help of the techniques described in this book, it is possible to design family of codes corresponding to respective communication systems which exhibit very good performance. In all the investigated cases, in fact, the performance is very close to the theoretical limit given by the mutual information of the corresponding channel.

Nonetheless, it is important to note that the path to the design of the "perfect" communication system is still unfinished. Several factors should be

M. Franceschini, G. Ferrari and R. Raheli, *LDPC Coded Modulations*,
DOI: 10.1007/978-3-540-69457-1_8,

taken into account and two of them may be a concern in practical scenarios.

The first and most important factor is the system complexity. In a relatively slow communication system, e.g., with transmission rate below 10Mbit/s, the technology available today allows to implement an LDPC coded modulation schemes by means of standard general purpose digital signal processors. This obviously guarantees a significant flexibility, which enables:

- the use of codes with a non-optimized structure;
- the use of long codewords;
- high precision arithmetics.

In a high-speed and, possibly, low-latency scenario, properly designed encoding and decoding schemes become a necessity and the use of generic unstructured LDPC codes would result in a prohibitive cost both for storing the code structure itself and for implementing the required ad-hoc interconnection in the LDPC coded modulation transmitter and receiver. Therefore, high-speed, low-latency systems pose challenging tasks such as:

- the design of highly structured and powerful LDPC codes for the LDPC coded modulation scheme of interest;
- the need to devise iterative message passing algorithm that guarantee good convergence properties even though used with low precision messages;
- the design of low complexity soft-input soft-output (SISO) modules for LDPC coded modulations; such a low complexity SISO algorithm for LDPC coded modulations is not necessarily a good SISO algorithm for uncoded modulation.

The second important factor that may be a concern is the sub-optimality of the considered encoding and decoding structures. The discussed analysis methods based on extrinsic information transfer (EXIT) charts give a useful estimate of the performance attainable with a given LDPC coded modulation system. This, however, does not guarantee that the considered scheme can always achieve the channel capacity.

Nevertheless, LDPC coded modulations represent a practical way of achieving good performance in a wide variety of channels with the currently available technology and may be regarded as a medium term flexible intermediate step toward yet-to-come low-complexity capacity-achieving communication schemes.

Bibliography

[1] T. M. Cover and J. A. Thomas, *Elements of Information Theory.* New York, NY, USA: John Wiley & Sons, Inc., 1991.

[2] D. Arnold, H.-A. Loeliger, P. O. Vontobel, A. Kavcic, and W. Zeng, "Simulation-based computation of information rates for channels with memory," *IEEE Trans. Inform. Theory*, vol. 52, no. 8, pp. 3498–3508, August 2006.

[3] R. Gallager, *Information Theory and Reliable Communication.* New York, NY, USA: John Wiley & Sons, 1968.

[4] J. G. Proakis, *Digital Communications,* 4th Edition. New York, NY, USA: McGraw-Hill, 2001.

[5] S. G. Wilson, *Digital Modulation and Coding.* Upper Saddle River, NJ, USA: Prentice-Hall, 1996.

[6] S. Lin and D. J. Costello, Jr., *Error Control Coding: Fundamentals and Applications.* Englewood Cliffs, NJ, USA: Prentice-Hall, 1984.

[7] C. Berrou, A. Glavieux, and P. Thitmajshima, "Near Shannon limit error-correcting coding and decoding: turbo-codes," in *Proc. IEEE Intern. Conf. on Commun.* (ICC), Geneva, Switzerland, May 1993, pp. 1064–1070.

[8] C. Berrou and A. Glavieux, "Near optimum error correcting coding and decoding: turbo-codes," *IEEE Trans. Commun.*, vol. 44, no. 10, pp. 1261–1271, October 1996.

[9] R. G. Gallager, *Low-Density Parity-Check Codes.* Cambridge, MA, USA: MIT Press, 1963.

[10] M. G. Luby, M. Mitzenmacher, M. A. Shokrollahi, D. A. Spielman, and V. Stemann, "Efficient erasure correcting codes," *IEEE Trans. Inform. Theory*, vol. 47, no. 2, pp. 569–584, February 2001.

[11] D. J. C. MacKay and R. M. Neal, "Near Shannon limit performance of low density parity check codes," *IEE Electronics Letters*, vol. 32, no. 18, pp. 1645–1646, 29th of August 1996.

[12] D. J. C. MacKay, "Good error correcting codes based on very sparse matrices," *IEEE Trans. Inform. Theory*, vol. 45, no. 2, pp. 399–431, March 1999.

[13] T. Richardson, A. Shokrollahi, and R. Urbanke, "Design of capacity-approaching irregular low-density parity check codes," *IEEE Trans. Inform. Theory*, vol. 47, no. 2, pp. 619–637, February 2001.

[14] A. J. Viterbi, "Error bounds for convolutional codes and an asymptotically optimum decoding algorithm," *IEEE Trans. Inform. Theory*, vol. 13, no. 2, pp. 260–269, April 1967.

[15] G. D. Forney, Jr., "The Viterbi algorithm," *Proc. IEEE*, vol. 61, no. 3, pp. 268–278, March 1973.

[16] J. Hagenauer and P. Hoeher, "A Viterbi algorithm with soft-decision outputs and its applications," in *Proc. IEEE Global Telecommun. Conf.* (GLOBECOM), Dallas, TX, USA, November 1989, pp. 1680–1686.

[17] L. R. Bahl, J. Cocke, F. Jelinek, and J. Raviv, "Optimal decoding of linear codes for minimizing symbol error rate," *IEEE Trans. Inform. Theory*, vol. 20, no. 2, pp. 284–287, March 1974.

[18] R. E. Blahut, *Principles and Practice of Information Theory.* Reading, MA, USA: Addison-Wesley, 1987.

[19] D. J. C. MacKay, *Information Theory, Inference & Learning Algorithms.* Cambridge, UK: Cambridge University Press, 2002.

[20] C. Shannon, "A mathematical theory of communication," *Bell System Tech. J.*, vol. 27, pp. 379–423, July 1948.

[21] A. J. Viterbi and J. K. Omura, *Principles of Digital Communication and Coding.* New York, NY, USA: McGraw-Hill, 1979.

[22] R. W. Chang and J. C. Hancock, "On receiver structures for channels having memory," *IEEE Trans. Inform. Theory*, vol. 12, no. 4, pp. 463–468, October 1966.

[23] K. Abend and B. D. Fritchman, "Statistical detection for communication channels with intersymbol interference," *Proc. IEEE*, vol. 58, no. 5, pp. 779–785, May 1970.

[24] P. L. McAdam, L. R. Welch, and C. L. Weber, "M.A.P. bit decoding of convolutional codes," in *Proc. IEEE Symposium on Information Theory* (ISIT), Asilomar, CA, USA, January 1972.

[25] G. Ferrari, G. Colavolpe, and R. Raheli, *Detection Algorithms for Wireless Communications*. Chichester, West Sussex, UK: John Wiley & Sons, 2004.

[26] ——, "A unified framework for finite-memory detection," *IEEE J. Select. Areas Commun.*, vol. 23, no. 9, pp. 1697–1706, September 2005.

[27] G. Ungerboeck, "Channel coding with multilevel/phase signals," *IEEE Trans. Inform. Theory*, vol. 28, no. 1, pp. 55–67, January 1982.

[28] J. Anderson, T. Aulin, and C.-E. Sundberg, *Digital Phase Modulation*. New York, NY, USA: Plenum Press, 1986.

[29] R. G. Gallager, "Low density parity check codes," *IEEE Trans. Inform. Theory*, vol. 8, no. 1, pp. 21–28, January 1962.

[30] C. Douillard, M. Jezequel, C. Berrou, A. Picart, P. Didier, and A. Glavieux, "Iterative correction of intersymbol interference: turbo-equalization," *European Trans. Telecommun.*, vol. 6, no. 5, pp. 507–511, September/October 1995.

[31] R. M. Tanner, "A recursive approach to low complexity codes," *IEEE Trans. Inform. Theory*, vol. 27, no. 5, pp. 533–547, September 1981.

[32] M. G. Luby, M. Mitzenmacher, M. A. Shokrollahi, and D. A. Spielman, "Improved low-density parity-check codes using irregular graphs," in *Proc. IEEE Symposium on Information Theory* (ISIT), Cambridge, MA, USA, August 1998, p. 117.

[33] S.-Y. Chung, G. D. Forney, T. J. Richardson, and R. L. Urbanke, "On the design of low-density parity-check codes within 0.0045 dB of the

Shannon limit," *IEEE Commun. Lett.*, vol. 5, no. 2, pp. 58–60, February 2001.

[34] T. Richardson and R. Urbanke, "The capacity of low density parity check codes under message passing decoding," *IEEE Trans. Inform. Theory*, vol. 47, no. 2, pp. 599–618, February 2001.

[35] D. J. C. MacKay, "Good error-correcting codes based on very sparse matrices," *IEE Electronics Letters*, vol. 33, pp. 457–458, March 1997.

[36] J. Pearl, *Probabilistic Reasoning in Intelligent Systems: Networks of Plausible Inference.* San Francisco, CA, USA: Morgan Kaufmann, 1988.

[37] F. R. Kschischang, B. J. Frey, and H.-A. Loeliger, "Factor graphs and the sum-product algorithm," *IEEE Trans. Inform. Theory*, vol. 47, no. 2, pp. 498–519, February 2001.

[38] H. Xiao and A. H. Banihashemi, "Graph-based message-passing schedules for decoding LDPC codes," *IEEE Trans. Commun.*, vol. 53, no. 12, pp. 2098–2105, December 2004.

[39] A. Tarable, S. Benedetto, and G. Montorsi, "Mapping interleaving laws to parallel turbo and LDPC decoder architectures," *IEEE Trans. Inform. Theory*, vol. 50, no. 9, pp. 2002–2009, September 2004.

[40] K. Gunnam, G. Choi, and M. Yeary, "An LDPC decoding schedule for memory access reduction," in *Proc. IEEE Intern. Conf. on Acoustics, Speech, and Signal Proc.* (ICASSP), Montreal, Canada, May 2004, pp. V–173–6 vol.5.

[41] D. J. C. MacKay, "Gallager codes – recent results," in *Coding, Communications and Broadcasting*, M. D. P. Farrell and B. Honary, Eds. Baldock, Hertfordshire, England: Research Studies Press, 2000, pp. 139–150.

[42] J. Chen and M. Fossorier, "Near optimum universal belief propagation based decoding of low-density parity-check codes," *IEEE Trans. Commun.*, vol. 50, no. 3, pp. 406–414, March 2002.

[43] J. Chen, V. Dholakia, E. Eleftheriou, M. Fossorier, and X.-Y. Hu, "Reduced-complexity decoding of ldpc codes," *IEEE Trans. Commun.*, vol. 53, no. 8, pp. 1288–1299, August 2005.

[44] J. Feldman, "Decoding error-correcting codes via linear programming," Ph.D. dissertation, Massachusetts Institute of Technology, 2003.

[45] J. Feldman, M. J. Wainwright, and D. R. Karger, "Using linear programming to decode binary linear codes," *IEEE Trans. Inform. Theory*, vol. 51, no. 3, pp. 954–972, March 2005.

[46] K. M. Chugg, A. Anastasopoulos, and X. Chen, *Iterative Detection: Adaptivity, Complexity Reduction, and Applications.* Reading, MA, USA: Kluwer Academic Publishers, 2001.

[47] N. Miladinovic and M. Fossorier, "Improved bit-flipping decoding of low-density parity-check codes," *IEEE Trans. Inform. Theory*, vol. 51, no. 4, pp. 1594–1606, April 2005.

[48] X. Hu, E. Eleftheriou, and D. Arnold, "Progressive edge-growth Tanner graphs," in *Proc. IEEE Global Telecommun. Conf.* (GLOBECOM), San Antonio, TX, USA, November 2001, pp. 995–1001.

[49] ——, "Regular and irregular progressive edge-growth tanner graphs," *IEEE Trans. Inform. Theory*, vol. 51, no. 1, pp. 386–398, January 2005.

[50] L. Chen, J. Xu, I. Djurdjevic, and S. Lin, "Near-shannon-limit quasi-cyclic low-density parity-check codes," *IEEE Trans. Commun.*, vol. 52, no. 7, pp. 1038–1042, July 2004.

[51] M. Fossorier, "Quasicyclic low-density parity-check codes from circulant permutation matrices," *IEEE Trans. Inform. Theory*, vol. 50, no. 8, pp. 1788–1793, August 2004.

[52] Y. Kou, S. Lin, and M. Fossorier, "Low-density parity-check codes based on finite geometries: a rediscovery and new results," *IEEE Trans. Inform. Theory*, vol. 47, no. 7, pp. 2711–2736, November 2001.

[53] G. Colavolpe, "Design and performance of turbo Gallager codes," *IEEE Trans. Commun.*, vol. 52, no. 11, pp. 1901–1908, November 2004.

[54] E. Eleftheriou and S. Olcer, "Low-density parity-check codes for digital subscriber lines," in *Proc. IEEE International Conf. on Commun.* (ICC), New York, NY, USA, April 2002, pp. 1752–1757.

[55] T. Richardson and R. Urbanke, "Efficient encoding of low-density parity-check codes," *IEEE Trans. Inform. Theory*, vol. 47, no. 2, pp. 638–656, February 2001.

[56] A. Papoulis, *Probability, Random Variables and Stochastic Processes.* New York, NY, USA: McGraw-Hill, 1991.

[57] G. S. Fishman, *Monte Carlo. Concepts, Algorithms and Applications.* New York, NY, USA: Springer, 1996.

[58] T. Richardson and R. Urbanke, *Modern Coding Theory.* Cambridge, UK: Cambridge University Press, 2008.

[59] D. Divsalar, S. Dolinar, and F. Pollara, "Iterative turbo decoder analysis based on density evolution," *IEEE J. Select. Areas Commun.*, vol. 19, no. 5, pp. 891–907, May 2001.

[60] S. ten Brink, "Convergence of iterative decoding," *IEE Electronics Letters*, vol. 35, pp. 1117–1119, 24th June 1999.

[61] ——, "Convergence behavior of iteratively decoded parallel concatenated codes," *IEEE Trans. Commun.*, vol. 49, no. 10, pp. 1727 – 1737, October 2001.

[62] I. Sutskover, S. Shamai, and J. Ziv, "Extremes of information combining," in *Proc. Allerton Conf. on Commun., Control, and Computing*, Monticello, IL, USA, October 2003.

[63] I. Land, S. Huettinger, P. A. Hoeher, and J. Huber, "Bounds on information combining," *IEEE Trans. Inform. Theory*, vol. 51, no. 2, pp. 612–619, February 2005.

[64] I. Sutskover, S. Shamai, and J. Ziv, "Extremes of information combining," *IEEE Trans. Inform. Theory*, vol. 51, no. 4, pp. 1313–1325, April 2005.

[65] S. Benedetto, D. Divsalar, G. Montorsi, and F. Pollara, "A soft-input soft-output APP module for iterative decoding of concatenated codes," *IEEE Commun. Lett.*, vol. 1, no. 1, pp. 22–24, January 1997.

[66] ——, "Soft-Input Soft-Output modules for the construction and distributed iterative decoding of code networks," *European Trans. Telecommun.*, vol. 9, no. 2, pp. 155–172, March/April 1998.

[67] P. Hoeher and J. Lodge, ""Turbo DPSK": Iterative differential PSK demodulation and channel decoding," *IEEE Trans. Commun.*, vol. 47, no. 6, pp. 837–843, June 1999.

[68] K. R. Narayanan and G. L. Stüber, "A serial concatenation approach in iterative demodulation and decoding," *IEEE Trans. Commun.*, vol. 47, no. 7, pp. 956–961, July 1999.

[69] S. ten Brink, G. Kramer, and A. Ashikhmin, "Design of low-density parity-check codes for modulation and detection," *IEEE Trans. Commun.*, vol. 52, no. 4, pp. 670–678, April 2004.

[70] E. Sharon, A. Ashikhmin, and S. Litsyn, "Analysis of low-density parity-check codes based on EXIT functions," *IEEE Trans. Commun.*, vol. 54, no. 8, pp. 1407–1414, August 2006.

[71] ——, "EXIT functions for binary input memoryless symmetric channels," *IEEE Trans. Commun.*, vol. 54, no. 7, pp. 1207–1214, July 2006.

[72] X. Li and J. A. Ritcey, "Bit-interleaved coded modulation with iterative decoding," *IEEE Commun. Lett.*, vol. 1, no. 6, pp. 169–171, November 1997.

[73] ——, "Bit-interleaved coded modulation with iterative decoding using soft feedback," *IEE Electronics Letters*, vol. 34, no. 10, pp. 942–943, May 1998.

[74] ——, "Trellis-coded modulation with bit interleaving and iterative decoding," *IEEE J. Select. Areas Commun.*, vol. 17, no. 4, pp. 715–724, April 1999.

[75] Y. Huang and J. A. Ritcey, "Exit chart analysis of bicm-id over awgn channels with snr mismatch," *IEEE Commun. Lett.*, vol. 8, no. 8, pp. 532–534, August 2004.

[76] M. Ardakani and F. R. Kschischang, "Designing irregular LDPC codes using EXIT charts based on message error rate," in *Proc. IEEE Symposium on Information Theory* (ISIT), Lausanne, Switzerland, June 2002, p. 454.

[77] ——, "A more accurate one-dimensional analysis and design of irregular LDPC codes," *IEEE Trans. Commun.*, vol. 52, no. 12, pp. 2106–2114, December 2004.

[78] M. Franceschini, G. Ferrari, and R. Raheli, "LDPC-coded modulations: performance bounds and a novel design criterion," in *Proc. Intern. Symp. on Turbo Codes & Relat. Topics*, Munich, Germany, April 2006.

[79] M. Franceschini, G. Ferrari, R. Raheli, and A. Curtoni, "Serial concatenation of LDPC codes and differential modulations," *IEEE J. Select. Areas Commun.*, vol. 23, no. 9, pp. 1758–1768, September 2005.

[80] E. Zehavi, "8-PSK trellis codes for a Rayleigh channel," *IEEE Trans. Commun.*, vol. 40, no. 5, pp. 873–884, May 1992.

[81] G. Caire, G. Taricco, and E. Biglieri, "Bit-interleaved coded modulation," *IEEE Trans. Inform. Theory*, vol. 44, no. 3, pp. 927–946, May 1998.

[82] A. P. Worthen and W. E. Stark, "Unified design of iterative receivers using factor graphs," *IEEE Trans. Inform. Theory*, vol. 47, no. 2, pp. 843–849, February 2001.

[83] G. Colavolpe, G. Ferrari, and R. Raheli, "Extrinsic information in iterative decoding: a unified view," *IEEE Trans. Commun.*, vol. 49, no. 12, pp. 2088–2094, December 2001.

[84] M. K. Simon, S. M. Hinedi, and W. C. Lindsey, *Digital Communication Techniques.* Englewood Cliffs, NJ, USA: Prentice-Hall, 1995.

[85] U. Mengali and A. N. D'Andrea, *Synchronization Techniques for Digital Receivers (Applications of Communications Theory).* New York, NY, USA: Plenum Press, 1997.

[86] M. K. Simon and D. Divsalar, "Multiple symbol differential detection of MPSK," *IEEE Trans. Commun.*, vol. 38, no. 3, pp. 300–308, March 1990.

[87] G. Colavolpe and R. Raheli, "Noncoherent sequence detection," *IEEE Trans. Commun.*, vol. 47, no. 9, pp. 1376–1385, September 1999.

[88] N. Noels, C. Herzet, A. Dejonghe, V. Lottici, H. Steendam, M. Moeneclaey, M. Luise, and L. Vandendorpe, "Turbo synchronization: an EM algorithm interpretation," in *Proc. IEEE Intern. Conf. on Commun.* (ICC), Anchorage, AK, USA, June 2003, pp. 2933–2937.

[89] M. Feder and N. Merhav, "Relations between entropy and error probability," *IEEE Trans. Inform. Theory*, vol. 40, no. 1, pp. 259–266, January 1994.

[90] M. Tüchler, "Design of serially concatenated systems depending on the block length," *IEEE Trans. Commun.*, vol. 52, no. 2, pp. 209 – 218, February 2004.

[91] R. Storn and K. Price, "Differential evolution–a simple and efficient heuristic adaptive scheme for global optimization over continuous spaces," *J. Global Optimization*, vol. 11, pp. 341–359, 1997.

[92] M. Franceschini, G. Ferrari, and R. Raheli, "EXIT chart-based design of LDPC codes for inter-symbol interference channels," in *Proc. IST Mobile Summit*, Dresden, Germany, June 2005.

[93] J. Huber, I. Land, and P. A. Hoeher, "Bounds on information combining," *IEEE Trans. Inform. Theory*, vol. 51, no. 2, pp. 612–619, February 2005.

[94] A. Ashikhmin, G. Kramer, and S. ten Brink, "Extrinsic information transfer functions: model and erasure channel properties," *IEEE Trans. Inform. Theory*, vol. 50, no. 11, pp. 2657–2673, November 2004.

[95] A. Kavćić, X. Ma, and M. Mitzenmacher, "Binary intersymbol interference channels: Gallager codes, density evolution, and code performance bounds," *IEEE Trans. Inform. Theory*, vol. 49, no. 7, pp. 1636–1652, July 2003.

[96] X. Ma and E. Yang, "Low-density parity-check codes with fast decoding convergence speed," in *Proc. IEEE Symposium on Information Theory* (ISIT), Chicago, IL, USA, July 2004, p. 274.

[97] R. Urbanke, "http://lthcwww.epfl.ch/research/ldpcopt/," web site.

[98] N. Varnica and A. Kavćić, "Optimized low-density parity-check codes for partial response channels," *IEEE Commun. Lett.*, vol. 7, no. 4, pp. 168–170, April 2003.

[99] A. Thangaraj and S. McLaughlin, "Thresholds and scheduling for LDPC-coded partial response channels," *IEEE Trans. Magn.*, vol. 38, no. 5, pp. 2307–2309, September 2002.

[100] C. Jones, T. Tian, A. Matache, R. Wesel, and J. Villasenor, "Robustness of LDPC codes on periodic fading channels," in *Proc. IEEE Global Telecommun. Conf.* (GLOBECOM), vol. 2, Taipei, Taiwan, November 2002, pp. 1284–1288.

[101] C. Jones, A. Matache, T. Tian, J. Villasenor, and R. Wesel, "The universality of LDPC codes on wireless channels," in *Proc. IEEE Military Comm. Conf.* (MILCOM), vol. 1, Boston, MA, USA, October 2003, pp. 440–445.

[102] F. Peng, W. E. Ryan, and R. D. Wesel, "Surrogate-channel design of universal LDPC codes," *IEEE Commun. Lett.*, vol. 10, no. 6, pp. 480–482, June 2006.

[103] H. Imai and S. Hirakawa, "A new multilevel coding method using error correcting codes," *IEEE Trans. Inform. Theory*, vol. 23, no. 3, pp. 371–377, May 1977.

[104] A. R. Calderbank, "Multilevel codes and multistage decoding," *IEEE Trans. Commun.*, vol. 37, no. 3, pp. 222–229, March 1989.

[105] U. Wachsmann, R. F. H. Fischer, and J. B. Huber, "Multilevel codes: theoretical concepts and practical design rules," *IEEE Trans. Inform. Theory*, vol. 45, no. 5, pp. 1361–1391, July 1999.

[106] I. Land, P. A. Hoeher, and J. Huber, "Analytical derivation of exit charts for simple block codes and for LDPC codes using information combining," in *Proc. European Signal Proc. Conf.* (EUSIPCO), Vienna, Austria, September 2004, pp. 1561–1564.

[107] ——, "Bounds on information combining for parity-check equations," in *International Zurich Seminar on Commun.* (ZTS'04), Zurich, Switzerland, September 2004, pp. 68 – 71.

[108] M. Franceschini, G. Ferrari, and R. Raheli, "Detection by multiple trellises," vol. 27, 2009, to appear.

[109] C. Rong-Rong, R. Koetter, U. Madhow, and D. Agrawal, "Joint noncoherent demodulation and decoding for the block fading channel: a practical framework for approaching shannon capacity," *IEEE Trans. Commun.*, vol. 51, no. 10, pp. 1676 – 1689, October 2003.

[110] M. Peleg, S. Shamai (Shitz), and S. Galán, "Iterative decoding for coded noncoherent MPSK communications over phase-noisy AWGN channel," *IEE Proceedings-Commun.*, vol. 147, no. 2, pp. 87–95, April 2000.

[111] G. Colavolpe and R. Raheli, "Theoretical analysis and performance limits of noncoherent sequence detection of coded PSK," *IEEE Trans. Inform. Theory*, vol. 46, no. 4, pp. 1483–1494, July 2000.

[112] ——, "The capacity of the noncoherent channel," *European Trans. Telecommun.*, vol. 12, no. 4, pp. 289–296, July/August 2001.

[113] M. Peleg and S. S. (Shitz), "On the capacity of the blockwise incoherent MPSK channel," *IEEE Trans. Commun.*, vol. 46, no. 5, pp. 603–609, May 1998.

[114] O. Macchi and L. Scharf, "A dynamic programming algorithm for phase estimation and data decoding on random phase channels," *IEEE Trans. Inform. Theory*, vol. 27, no. 5, pp. 581–595, September 1981.

[115] P. Yahampath and M. Pawlak, "Vector quantisation for finite-state Markov channels and application to wireless communications," *European Trans. Telecommun.*, vol. 18, no. 2, pp. 327–342, February 2007.

[116] R. H. Clarke, "A statistical theory of mobile radio reception," *Bell System Tech. J.*, vol. 47, no. 6, pp. 957–1000, August 1968.

[117] W. C. Jakes, *Microwave Mobile Communications.* New York, NY, USA: John Wiley & Sons, 1974.

[118] W. Turin and R. V. Nobelen, "Hidden markov modeling of flat fading channels," *IEEE J. Select. Areas Commun.*, vol. 16, no. 9, pp. 1809–1817, December 1998.

[119] C. C. Tan and N. C. Beaulieu, "On first-order Markov modeling for the Rayleigh fading channel," *IEEE Trans. Commun.*, vol. 48, no. 12, pp. 2032–2040, December 2000.

List of Acronyms

ACS	Add-Compare-Select
APP	A-Posteriori Probability
APSK	Amplitude and Phase Shift Keying
ASK	Amplitude-Shift Keying
AWGN	Additive White Gaussian Noise
BAC	Binary Asymmetric Channel
BCH	Bose-Chaudhuri-Hocquenghem
BCJR	Bahl, Cocke, Jelinek, Raviv
BEC	Binary Erasure Channel
BEP	Bit Error Probability
BER	Bit Error Rate
BI	Binary-Input
BIAWGN	Binary-Input Additive White Gaussian Noise
BICM	Bit-Interleaved Coded Modulation
BICM-ID	BICM with Iterative Detection
BP	Belief Propagation
BPSK	Binary Phase Shift Keying
BSC	Binary Symmetric Channel
CM	Coded Modulation
CM-SISO	Coded Modulator Soft-Input Soft-Output module
CND	Check Node Detector
CPM	Continuous Phase Modulation
CSI	Channel State Information
DD	Differential Detector
DE	Differentially Encoded
DE-PSK	Differentially-Encoded Phase Shift Keying
DE-QAM	Differentially-Encoded Quadrature Amplitude Modulation
DPI	Data Processing Inequality
DPSK	Differential Phase Shift Keying
ECC	Error Correcting Code

EXIT	EXtrinsic Information Transfer
FB	Forward-Backward
FER	Frame Error Rate
FMC	Finite Memory Condition
FSM	Finite State Machine
i.i.d.	independent and identically distributed
IR	Information Rate
ISI	Inter-Symbol Interference
LDPC	Low Density Parity Check
LLR	Logarithmic Likelihood Ratio
LP	Linear Programming
MAP	Maximum A Posteriori
MI	Mutual Information
MIMO	Multiple-Input Multiple-Output
ML	Maximum Likelihood
MLC	Multi-Level Coding
MSD	Multi-Stage Decoding
OOK	On-Off Keying
PAM	Pulse Amplitude Modulation
PEG	Progressive Edge Growth
PCCC	Parallel Concatenated Convolutional Code
pdf	probability density function
pmf	probability mass function
PRC	Partial Response Channel
PRNG	Pseudo-Random Number Generator
PSK	Phase Shift Keying
QAM	Quadrature Amplitude Modulation
QPSK	Quaternary Phase Shift Keying
RNG	Random Number Generator
RX	Receiver
RV	Random Variable
SER	Symbol Error Rate
SISO	Soft-Input Soft-Output
SNR	Signal-to-Noise Ratio
SOVA	Soft-Output Viterbi Algorithm
SP	Sum-Product algorithm
TCM	Trellis Coded Modulation
TX	Transmitter
VA	Viterbi Algorithm
VND	Variable Node Detector
ZC	Z Channel

Index

algorithm
- Bahl-Cocke-Jelinek-Raviv (BCJR), 9
- belief propagation (BP), 43
- forward-backward (FB), 25
- min-sum, 49
- progressive edge growth (PEG), 53
- soft output Viterbi (SOVA), 9
- sum-product (SP), 42
- Viterbi, 9, 21

average variable node degree, 117

BICM, 83

bipartite graph, 38
- short cycles, 135, 153

bit error rate (BER), 4
- lower bound, 94
- optimization criterion, 110
- upper bound, 94

channel
- binary asymmetric, 133
- binary input AWGN, 133
- binary symmetric, 133
- erasure, 133
- fading, 163, 172
- memoryless, 130, 151
- phase-uncertain, 163, 170
- Z, 133

channel coding, 4

code
- block, 7
  - linear, 8
- convolutional (stream), 9
- error correcting, 7
- low-density parity-check (LDPC), 37
  - decoding, 42
  - encoding, 56
  - practical design, 53
  - statistical description, 40
- structured, 58, 83
- turbo, 33

communication system, 3

computational cost, 156

convergence threshold, 100, 135

decoding
- convergence, 92, 103
- iterations, 90, 112, 114
  - final, 101

degree distribution, 97

density evolution, 68, 95

detection, 19
- channels with memory, 30
- noncoherent, 159

detection and decoding
- graph-based, 85
- iterative, 33
- joint, 32
- separate, 33
- turbo approach, 86

detector
- check node (CND), 49, 89

variable node (VND), 49, 89
differentially encoded
phase shift keying (PSK), 107, 148
QAM, 157

extrinsic information, 89
extrinsic information transfer (EXIT) chart, 71
approximation, 91
decoding trajectory, 91, 102

forward backward algorithm, 9

girth, 54
graph connectivity, 117

information theory
entropy, 10
information theory
basics, 10
channel capacity, 12
information rate (IR), 11
interleaver, 83
intersymbol interference (ISI) channel, 108, 121

LDPC code
irregular, 41
library, 137
regular, 120, 144
LDPC coded modulation, 87
design, 95
linear block codes, 8

mapping
Gray, 121, 123, 126
Markov chain, 164
time-varying, 164
metrics
normalization, 167
modulation, 5, 6
Monte Carlo simulation, 62
multi-level coding, 130, 137
multi-level coding (MLC)
design algorithm, 141
multi-stage decoding (MSD), 138

node
check, 38
variable, 38

performance analysis, 61
pseudo-random number generator (PRNG), 62

soft-input soft-output block, 8
sufficient statistics, 131

theoretical limits, 4
turbo
codes, 33
detection, 35
equalization, 36

MIX
Papier aus verantwortungsvollen Quellen
Paper from responsible sources
FSC® C105338

If you have any concerns about our products,
you can contact us on
ProductSafety@springernature.com

In case Publisher is established outside the EU,
the EU authorized representative is:
Springer Nature Customer Service Center GmbH
Europaplatz 3, 69115 Heidelberg, Germany

Printed by Libri Plureos GmbH
in Hamburg, Germany